ROUTLEDGE LIBRAR
SOCIAL AND CULTURA

Volume 11

THE SCOPE OF GEOGRAPHY

THE SCOPE OF GEOGRAPHY
Third Edition

RHOADS MURPHEY

Routledge
Taylor & Francis Group

LONDON AND NEW YORK

First published 1969 by Rand McNally

Published 2016 by Routledge
Second edition published in 1973 by Rand McNally
Third edition published in 1982 in the USA by Methuen & Co.

This edition first published in 2014
by Routledge
2 Park Square, Milton Park, Abingdon, Oxon OX14 4RN

and by Routledge
711 Third Avenue, New York, NY 10017, USA

Routledge is an imprint of the Taylor & Francis Group, an informa business

First issued in paperback 2015

British Library Cataloguing in Publication Data
A catalogue record for this book is available from the British Library

ISBN 978-0-415-83447-6 (Set)
eISBN 978-1-315-84860-0 (Set)
ISBN 978-0-415-73347-2 (hbk) (Volume 11)
ISBN 978-1-138-98995-5 (pbk) (Volume 11)
ISBN 978-1-315-84833-4 (ebk) (Volume 11)

Publisher's Note
The publisher has gone to great lengths to ensure the quality of this reprint but points out that some imperfections in the original copies may be apparent.

Disclaimer
The publisher has made every effort to trace copyright holders and would welcome correspondence from those they have been unable to trace.

Rhoads Murphey

THE SCOPE OF GEOGRAPHY

Third Edition

Methuen

NEW YORK AND LONDON

First published in 1969 by Rand McNally
Second edition published in 1973 by Rand McNally
Third edition published in the USA by
Methuen & Co.
in association with Methuen, Inc.
733 Third Avenue, New York, NY 10017

Third edition first published in 1982 by
Methuen & Co. Ltd
11 New Fetter Lane, London EC4P 4EE

© 1969, 1973, 1982 Rhoads Murphey

Photoset by Nene Phototypesetters Ltd, Northampton
and printed in the United States of America

Library of Congress Cataloging in Publication Data

Murphey, Rhoads, 1919–
The scope of geography.
Includes bibliographies and index.
1. Geography.   I. Title.
G116.M87   1982   910   82–6471

ISBN 0–416–33410–5   AACR2

British Library Cataloguing in Publication Data

Murphey, Rhoads
The scope of geography.
1. Geography
I. Title
910   G128

ISBN 0–416–33410–5

CONTENTS

Introduction to the Third Edition   page ix

1	Geography and social science	1
2	The region	8
3	Spatial form and spatial interaction	23
4	Spatial systems	42
5	The environmental factor	57
6	The ecological imperative	71
7	Culture, nature, and climate	84
8	The workings of weather and climate	96
9	Vegetation and human settlement	114
10	Soils and agriculture	128
11	Water and people	147
12	Land forms and settlement	162
13	Resources and population growth	176
14	Energy and development	190
15	Resource use, industrialization, and the future	208
16	Cities and manufacturing: spatial concentration	229

Suggestions for further reading   245

Possible study/discussion questions   249

Index   253

FIGURES

1	Overlapping regional traits	page 13
2	Hypothetical spatial patterns of city hinterlands	29
3	City hinterlands in a hypothetical landscape	30
4	Simple interaction diagram	32
5	Sample spatial patterns	35
6	Projection comparisons	38
7	Latitude and longitude	40
8	A New Yorker's view of the United States	44
9a	Villages on a south Indian plain	47
9b	Standard Asian village pattern	47
10	Models of urban spatial structure	49
11	Hypothetical land use zones around a city	50
12	World temperature trends	92
13	Cyclonic precipitation	98
14	Convectional precipitation	99
15	Orographic precipitation	100
16	Generalized wind belts of the world	103
17	Earth-sun relationships	105
18	Generalized vegetation sequence	116–17
19	World lumber, pulp, and wood cut	122
20	Sample soil profile	132
21	Ground water occurrence	149
22	Hypothetical settlements along a river	159
23	World population, 1968 and 2000	184
24	World population growth, AD 600–2000	188
25	Urban-rural divisions by continents	229

MAPS

1	Sample regionalizations of the United States	page 18
2	Land use in California and Nevada	19
3	Timbuktu in West Africa	25
4	Part of southeastern Michigan, from topographic sheet	37
5	Origins of commuters to Chicago	46
6	Land use in Boston	51
7	Cities of the fall line	61
8	Switzerland: altitudes and travel routes	64
9	Temperature inversions in the United States	76
10	World population density	89
11	World climatic types	91
12	United States railroads	109
13	United States population density	110
14	World vegetation types	115
15	Soil erosion in the United States	124
16	World soil types	129
17	Tenancy in Europe	140
18	World rainfall	148
19	Land subsidence in California	150
20	Irrigation in the United States	153
21	World land form types	164
22	Made land and land use in Seattle	174
23	World per capita income	204
24	World life expectancy at birth	205
25	World infant mortality	206

26 Major world manufacturing areas   214
27 World coal, iron, and oil   216–17
28 Coal, iron, and population in Europe   220
29 Metropolitan tributary areas in the United States   240
30 The United States manufacturing belt   240

Table

1 World population 1650–1980   182

INTRODUCTION
TO THE
THIRD EDITION

This book has grown out of an earlier book originally called *An Intro-duction to Geography*. The two have diverged through subsequent editions and each now deals quite differently with the field. The original *Introduction*, now called *Patterns on the Earth* and published by Houghton Mifflin in hard cover, is designed primarily for courses in regional geography, or for courses which introduce the student to the study of the field through the analysis of world regions. *The Scope of Geography* aims instead to provide an overall systematic view of the discipline of human geography as of the 1980s. It has been and remains a rapidly changing field, and the present edition incorporates those changes, presenting geography as a dynamic entity which addresses itself to the evolving problems of the modern world.

This new edition also provides greatly expanded cartographic cover-age, with over twenty new maps, each keyed to the text and designed to enhance student understanding of the material and concepts introduced. Illustrative figures and diagrams have also been expanded and re-drawn, and several new ones added. In response both to changes within the field and to suggestions from many readers of earlier editions, the entire text has been re-worked, reorganized, and re-written, shortened and sim-plified in some sections, expanded in others, and with new material presented in every chapter. Revisions on nearly every page have cor-rected errors and made the material both current and clearer. Two entirely new chapters have been added, on "Energy and Development" and on "Cities and Manufacturing," reflecting changes in the field, in the world with which it deals, and consequently in the needs of students and instructors in introductory geography courses.

While no one book can hope to meet exactly all the needs of every course or of every instructor, it is hoped that this third edition is now well suited for introductory courses in human geography. I have taught such courses myself for nearly thirty years and have learned a great deal, from my students, from colleagues in many places, and from comments on earlier editions of this book. I am grateful to all of them, and especially to those readers whose conscientious and thoughtful suggestions have contributed so much to this new edition. It is a pleasure also to thank Kathy Price and Monica Robinson for their cheerful and skillful work under pressure in preparing many of the new maps and figures.

Finally, I hope that the book may be useful also in a wider sense, as a survey of a field whose recent evolution makes it even more essential to all who are concerned about the world in which we live.

Rhoads Murphey

January, 1982

1 GEOGRAPHY AND SOCIAL SCIENCE

"Geography" comes from a Greek word meaning literally "description of the earth." But modern geography is concerned with people as well as with the physical earth and with relationships and analysis as well as with description. Geographers analyze the physical landscape and examine relations between places in order to throw light on the pattern and nature of human society. They investigate the interrelationship which exists between people and their physical environment. They examine regional differences and attempt to account for them, pick out regional patterns, try to draw regional lines and to discern regional relationships. Geographers direct this study of the earth and its spatial framework, or the pattern of distribution of things on the earth's surface, toward a better understanding of the human world. They set people and their culture in the framework of the world which they inhabit.

The broad outlook which is necessary when dealing with the analysis of spatial patterns over the world and of the varying regions and human use of the earth involves the geographer with a great number and variety of facts. *By themselves, as separate individual pieces of information or simple descriptions of places and features, facts are less useful than if they can be related and where possible used to reach conclusions. Information is not the same as knowledge.* It is only when facts can be put together to form a pattern or to support a general statement of relationship, or when conclusions can be drawn, that progress is made toward understanding. Facts never "speak for themselves;" they have meaning only when they can be related or grouped into some sort of system. This book attempts to guide the student toward the use of certain facts by means of the body of ideas which are employed in the field of geography.

Geographic analysis

Each of the social sciences represents one approach among many to their common objective of knowledge about human society, and the distinctiveness of these various approaches is more a matter of what each is looking for than of a set of mutually exclusive facts, tools, or methods. In any given set of data geographers look for particular factors and relationships to which they are attracted by their particular attitude. The same material may also be examined by the historian, for example, or the economist. (Of course, the geographer may also make use of purely historical or economic data.) Broadly, in any data he examines, the geographer is looking for spatial form and spatial relations or, in other words, for patterns of distribution and interaction. One may say that all geography begins with a map. The other disciplines are less concerned with such patterns and are therefore less likely to discover them, although they may find the geographer's conclusions useful toward their own purposes.

Distinct areas of analysis explored by geography may be summarized under three main headings as follows:

1. The regional framework and the analysis of specific regions.
2. The distribution and relationships of people over the earth and the spatial aspects of human settlement and use of the earth.
3. The interrelationship between human society and the physical environment as part of the study of areal differences.

The reader may complain that "space" is a very broad or even ambiguous word. Indeed, all three of these headings might be lumped together as spatial study. The division of the earth into regions and the analysis of specific regions are both concerned with spatial form and distribution. Examination of differences in human society from place to place as they may be related to differences in the physical environment, and of differing landscapes as they reflect a changing human agency, are also concerned with earth space. Environmental factors such as mountains, rivers, or deserts affect relations between one place and another and thus also affect the nature and distribution of human activities, which may in turn transform or severely damage the physical environment. Geography is basically concerned with the patterns of human distribution over and use of the earth, why people use it differently in different places or at different times, and how this is related to the whole of their cultures and economies. These are broad, complex questions; many approaches in various fields may be used toward answering them. Geographers must depend not only on their own analyses but also on the approaches and findings of others. But the phenomena which they choose to investigate, those in which matters of spatial distribution or the physical environments and landscape or the region are involved, lend themselves to geographic analysis.

The geographer, like other social scientists, believes in cause and effect. The pattern, organization, and nature of human society are not haphazard. Their origins are seldom simple, and some of them can be discovered only in part or not at all. But this does not invalidate the attempt to discover them nor make it a waste of time when the attempt fails to produce definitive or measurable results. If we do not ask questions, we do not learn. And if we believe in the importance of causality, we must also believe in the applicability of principles or concepts. *Laws* is too extreme a term and may never be fully appropriate in geography, at least not in the sense that the chemist can use Boyle's law or the biologist can use the Mendelian laws of heredity. Human society is less regular than the physical world and it cannot yield the same kinds of results to the student of it. The theoretically ordered behavior of society is the result of almost infinite variables as causes. It cannot be understood or predicted as well as the behavior of individual people, let alone the behavior of physical matter. Our aim is to introduce a degree of order, to establish relationships, patterns, associations of factors. This is seldom easy. Because phenomena are associated or occur together in space or in time it does not necessarily mean that they are causally related. It may mean this, or it may not. Such an association is more likely to be part of a much more complex relationship among a great variety of *interacting* factors, each influenced by the presence of the others. No one of them alone is both necessary and sufficient to produce an observed result, but taken together they may form a causal association or a consistent pattern which may recur under similar conditions elsewhere, although it is extremely rare for it to occur in exactly the same way twice. So many different and interacting factors are involved in most aspects of human society that in each case the necessary set of circumstances which created it are almost never duplicated. But although each area, each historical period, and each human phenomenon is different, certain common features can be found. Without this kind of analysis, there is merely an accumulation of data, which by itself may often be meaningless. The student of society must be concerned with these common features and other complex structures of causality and must attempt to distinguish simple coincidence or association from cause and effect.

Science and social science

It is seldom possible in the social sciences to achieve the precise results frequently characteristic of the physical sciences. This lack of precision is not so much because the study of society, in its modern form, is comparatively recently developed as because much of its material does not easily lend itself either to accurate measurement or to controlled conditions of observation and experiment. Human society is vastly complex, always changing, always in motion. Almost nothing about it is constant, so that at each point in time we may attempt to examine it a different combination of factors, known or unknown, and a different pattern of effects confront us. We must distinguish qualitative changes as well as more easily measured quantitative changes, whether we are attempting to study a particular process or simply an aspect of society. But we must be careful, for even without regard to change, material which we may arbitrarily attempt to isolate for examination may be quite unrepresentative, however carefully it is chosen, since we are dealing not with inert matter, which usually behaves regularly, nor even with biological matter in which one red blood cell is like ten thousand others, but with something far less orderly or fixed, such as a city or a system of land use. Each specimen is likely to be unique. We may speak of examining it, but we cannot take it into a laboratory; we cannot manipulate it but can only observe it. And however small the problem may be, it is impossible, without distortion, to isolate any aspect of human society from its manifold interconnections with other aspects.

Some things about human society, and some aspects of some things, may be more or less precisely measurable under favorable circumstances, such as, for example, population totals for definite areas, or value of recorded foreign trade. But many things and many aspects of it are not, or are measurable only in small part or by rough approximation, such as, for example, the basis of political power, or regional differences in culture, or population totals and trade flows in many developing societies. It is clear that the student of society must depend upon more than quantitative methods, although quantitative analysis can be very helpful where it is appropriate. Geography, especially, by its very nature incorporates an enormous field of study, encompassing most of human experience. Geographers must depend on a *variety of methods* in their analyses and cannot limit their investigations either to strictly quantitative measurements or to purely subjective or intuitive judgments.

In many ways it would be preferable to avoid the word "science" in connection with the study of society and to divide all learning into *humanities* and *sciences*, the first concerned with the nature and works of people, the second with the nature of the physical world. Most systematic inquiry into the nature and works of people, however, is best undertaken by means of the logical method, or, as it is sometimes rather ambiguously called, the "scientific method." Logic is far from being the exclusive property of the physical sciences, and indeed it has been most highly developed in one of the humanities, philosophy. Rational analysis of any kind depends on logical inquiry. The physical sciences differ from the social sciences or the humanities mainly in the nature of their material. Experimentation is a major basis of the logical method since only thus can alternative hypotheses be tested until the most suitable one is established. But experimentation in the laboratory and test-tube sense is impossible in most analysis of society. Students of society must be content by and large to apply alternative explanations and conceptual systems to social situations as they exist and to see whether they have really explained all the facts. With the interlocking complexity, awkwardness, and volatility of this material, with the inability to isolate or manipulate it, and with the handicap that we are far from having all or perhaps even most of the facts about mankind, it is understandable that we cannot compete with the accuracy and finality of the physical scientist in dealing with most questions.

Actually, the physical scientists are themselves finding that the physical world is less determinate and conforms less to a regular fixed order than has previously been assumed. Many of the long-accepted "laws" of physics, for example, have been disproved, and many of the seeming regularities in nature have been shown to be misleading. In this sense, there are no "natural" laws, but only human-perceived regularities, patterns, or associations. Students of society are, however, still more dependent than physical scientists on qualitative rather than quantitative methods, and they must often make value judgments. This does not make the study of society less interesting or less important, but it does make it different from precise, determinate science. Literal minds which like to think in black-and-white terms or in terms of fixed "laws" are not well suited to theoretical science of any kind, and certainly not to the study of society. The student of society believes in the scientific or logical method, but in terms of his job this cannot be more precisely defined than *the careful use of facts, reason, and systematic thought, aimed at the discovery of patterns or associations.*

Prediction and causality

Prediction is a somewhat controversial subject among students of society. It would be foolish to maintain that prediction is either entirely impossible or undesirable. But it is plain enough, in terms of the foregoing discussion, that most students of society are not in a position to read the future with precision except in limited cases. Prediction is an important function, and may in certain cases be extremely valuable. But the social sciences, certainly geography, do not stand or fall on their ability to predict accurately, nor should that be their chief objective for the future. Their work is not invalidated because it does not deal with or produce laws of complete predictive value; it is more appropriate for most analysis of society to speak in terms of probabilities or recurrent patterns. But when social scientists are so far from laboratory conditions in dealing with the present or the past, it is not to be expected that they can do any better, if as well, for the future, when many of the assumptions they make and the factors underlying the patterns they observe in the present or the past may be radically changed. Human society is never constant, and although it is part of the scholar's function to pick out patterns of its inconstancy, prediction is only an incidental part of this function. We are concerned with patterns and associations in human society as we are able to observe it; if this analysis helps to suggest patterns of future development, so much the better. Perhaps in time, as we learn more, some of these suggestions may begin to have greater predictive value, and no one would be happier about that than the student of society. But prediction will not then become our main function. We still have more than enough important and useful work to do toward understanding human society as it is and as it has developed, which is the principal approach of this book.

It should not need arguing that any educated person should understand as much as possible of world society. An understanding of human distribution over the earth, the fragile interrelationship with the environment, the different uses to which people have put different parts of it, the cultures and economies they have created, and the spatial interrelations which exist between and have influenced these patterns – or in a word, geography – has a fundamental place in the equipment of an educated person.

In the search for consistent patterns and associations, the social scientist is seldom in a position to say, as the physical scientist often is, "If a is present, then b will follow." People are not that simple, nor do we know enough about human society or about the interlocking factors influencing it. We are more likely to be able to say, "If a, b, and c are present, then d, e, or f may also be associated with them, but not g, h, i, j, or k." We can usually eliminate irrelevant factors and can usually determine the set of factors or conditions among which to seek a solution, but not often much more. We must also assume, as a result of accumulated experience, that wholes or aspects of societies or economies can rarely be explained adequately in terms of a single factor. The very complexity of human society and its interrelations, which makes simple explanations inadmissible, also makes a continual temptation to cut through the confusion, have done with the if's and but's, and devise a simple direct idea or set of ideas which can consistently explain everything. There are many famous examples of this kind of fallacious analysis. Marxism, especially as it has been used by people less principled or objective than Marx, is an outstanding one. There are many theories of history open to the charge of single-factor oversimplification. In the field of geography, the single-factor heresy is represented by *environmental determinism*, which maintains that everything, or nearly everything, about mankind is directly traceable to some aspect of the physical environment.

The damage which these misguided enthusiasts do is very great but it is often obscured by the insights which they simultaneously create. They successfully dramatize an idea, usually a most important one, or develop a theory which throws new light on a major problem. The more useful and interpretive they make the idea the harder it is to make clear that far from simplifying the problem they are often confusing it further by attempting to disguise its complexity. This makes an adequate explanation more difficult, especially when it must compete with the dramatic appeal of a "simple" answer by a weary recitation of complex problems, complex intercausation, and complex qualified solutions. The proper student of society can proceed only on the assumption that all single-factor explanations are to be distrusted if not rejected, and that we must always preserve (1) an open mind about what may be important or relevant and

(2) a stock of multiple explanations, or a readiness to formulate new ones, for every situation which we investigate. We must try not to be burdened with preconceived theories into which we attempt to squeeze ill-fitting data, but to use data to construct and to prove or disprove theories which we are always prepared to modify or discard. Generalizations are not developed in order to construct a theoretical system as an end in itself. The goal is the fuller understanding of people and their works.

Progress towards such understanding is not well served by pursuing any set theory to the exclusion of others or by concentrating on the use of any one method. This error appeared in the discipline of geography at one stage in its development in the form of environmental determinism. In the present generation it is important to resist the assumption that all of human society can be explained, or predicted, by quantitative analysis, or that all of geography can be made an exact "science." But it is equally important for geographers not to ignore the question of causality. We must not merely gather and arrange data or write description; we must continually ask why? In this endeavor, multiple methods and multiple ideas are relevant and necessary for fruitful results.

2 THE REGION

The study of spatial form or of the spatial dimensions of human society must take account of distinct areal units and is therefore concerned with the concept of the region. The world may be divided into continents or countries or counties according to the existing more or less arbitrary lines. But there is a more important and more complex set of areal divisions to be made on the basis of units which contain distinct and internally consistent patterns of physical features or of human development.

Aspects of areal difference

Such division is often difficult. There are almost endless criteria on which to base the definition or delimitation of a given region. An area may be distinctive and internally consistent because it coincides with the distribution of a particular type of climate, land forms, or soil, or because it is characterized by a particular type of economic activity, occupational structure, land use, or language. Areal distinctions may also be drawn on the basis of administrative or commercial services or trade relations which cover a particular territory. One may describe and mark off on the map manufacturing regions, agricultural regions, climatic regions, land form regions, religious regions, or political regions, to mention only a few of the more obvious ones. Few of these will often coincide exactly with each other, although in many cases the various factors or characteristics which make an area distinct may co-vary in their spatial distribution. The city is a good illustration of this problem. A city's newspapers will cover one area, its trade another (or a variety of others), the plain or river valley in which the city lies another, the city's specialized services, such as bus routes or library circulation, another, and its political control still another, the last one seldom coinciding with the compact urban area of the city.

For a variety of reasons it is essential to pick out some degree of regional distinction and regional coherence from the jumbled mass of distribution and areal differences. This is necessary in order to understand any one area adequately, to distinguish it from other areas, and to see the basic spatial framework within which people live and which they have helped to create. Planning for the management and development of any area requires an understanding of the limits of the regional unit involved and of the region's distinctive components. Regional planning, in which development is planned for an area as a whole in all of its interrelated aspects, has come to be accepted as an essential method for promoting economic growth or areal development. It must rest on a knowledge not only of regional differences and their delimitation but of the multiplicity of interrelated factors, from climate to transport patterns to historical trends, which compose any regional unit. The businessman and the political administrator also need to delimit areas and to understand them for the efficient operation of their affairs and the subdivision or centralization of their activities. Trade areas and political boundaries are often matters of dispute, reflecting the complex problems involved, and the geographer attempts to provide some guidance. The study of political boundaries, what makes a good one or a bad one, where in particular cases the boundary should best be put, and what impact a political boundary may have on the area through which it runs, is a whole field in itself, within political geography, but it is also part of the geographer's more general regional function.

Lines, barriers, and transition zones

In order to make regional divisions intelligently, it is necessary first to understand as well as possible what gives the region its coherence and distinction, what its essential unifying or common qualities are – its *core characteristics* – and secondly to determine the points or zones where these qualities significantly diminish and are replaced by another set of qualities belonging presumably to another region. Usually there is a wide *zone of transition*, and the line is difficult to draw. Prominent physical features such as mountain ranges often coincide areally with the zone of transition from one set of regional characteristics to another. Mountains may act as barriers to economic and cultural movement or interchange, and they may also be responsible for a transition in the physical aspects of the landscape. But the effect on people of even pronounced features of the environment is seldom constant. If such features are used as regional lines, the line may become obsolete or awkward as human occupation changes on both sides of such a boundary or meets across it. Even at a given moment in time sharp cultural gradients on the map are rare, and even in physical terms it is seldom possible to draw a neat line between regions. Regional characteristics change more or less gradually over space, and within the transition area between one region and another there may be a wide zone of ambivalence. This may be true even where a major mountain range is used as a regional line. The sea makes a clearer delimitation, but its barrier character has been tremendously lessened by the modern revolution in communications and transport. The North Atlantic Community, for example, now expresses in many respects more genuine unity among its members than exists among the countries of Europe as a whole, although the Atlantic is as wide and deep as ever. Rivers may be physical barriers to an extent, but where they are navigable they are usually unifiers in the sense that settlements on both banks are likely to be dependent on river-borne trade and hence will be part of a single economic and cultural community. Unnavigable rivers may support in their valleys a more or less uniform pattern of agricultural land use, which may be dependent on river water for irrigation. River basins are usually physical and human regional units. Attempts to divide them or to use rivers as political boundaries, as is commonly done, are unlikely to be satisfactory.

Distinction, identity, and regionalism

Similar problems are involved in the drawing of almost any regional line, since regions tend to merge with each other over a wide zone of transition despite physical barriers. Any single line is thus not only arbitrary but likely to be seriously misleading as well as awkward, even though for most purposes and especially for political boundaries a single line must be used. Physical features, especially rivers, provide convenient lines and are therefore used despite the disadvantages. But for regional analysis, the basic problem is the determination of what gives any region its essential distinctive quality, and it is only on that basis that satisfactory regional lines can be drawn. Here we must be concerned with the physical framework, with economic, political, and social characteristics and functions, and also with the regional images and loyalties of the inhabitants. We are dealing in many ways with what psychologists call a *Gestalt*, a pattern of factors which tends to operate as a unit and which is more than the simple sum of its parts, as water is more than the separate totals of its hydrogen and oxygen components. Each aspect of an area exists in context, is interrelated with other aspects, and cannot be understood completely unless it is seen in its total context, nor can the area as a whole be understood unless all of its interrelated aspects are seen. The consciousness of this greater whole, of this areal pattern of characteristics, in the minds of the inhabitants is called "regionalism," or in the case of a national state, "nationalism." Separatism, or the desire for political autonomy, is an aspect of pronounced regionalism which is akin to nationalism but which may not insist on complete independence. Outside threats, pressures, or competition, real or imaginary, commonly strengthen regional consciousness and cohesion, but regionalism in basic terms results simply from the awareness of areal differences. People identify themselves with their own area and with its characteristics, including diets, dress, house styles, art forms, political attitudes, landscapes, seasons, and a host of other attributes which differ from region to region. Parents inculcate regional cultural traits in their children, take pride in their regional identity, and feel themselves to be different from and usually better than the people of other regions.

Such attitudes provide one of the clearest clues to the existence of regional entities and are particularly helpful when it comes to drawing regional lines. This may be done in some situations merely by asking the inhabitants to which cultural region they feel they belong. Regionalism is often especially pronounced in isolated or mountainous areas, where the physical distinction from surrounding areas is strong and where the culture and economy are likely to be distinctive also. But every area which can be distinguished from neighboring areas feels it, in differing degrees. In the United States this feeling was strong enough in the southern states to support a movement for independence as a separate sovereignty, and it remains strong, partly because of sentiment and loyalties, partly because of the geographical differences which fostered the feeling of distinctiveness in the first place.

Language is often used as a measure of regionalism and as a basis for delimiting regions. Of all aspects of culture, language may be the most important mark of identity as well as one of the clearest, and it often does tend to coincide areally with most other aspects of a regional *Gestalt*. But although it may in many cases be a badge of regional identity whose importance is emotional as well as practical, language is not a necessary or sufficient cause for regionalism or an adequate measure of a regional entity. The example of the American Confederacy suggests that regionalism may be very strong and that a true region may exist without a regional language. Such examples could easily be multiplied. Switzerland, on the other hand, suggests that regionalism on other grounds than language may be strong enough to thrive despite language differences within the region. Factors other than language are clearly important in promoting regionalism and in creating a regional entity. Improvements in communication and the increase of interaction in the last several centuries have also been accompanied by an increase rather than a decrease in regional consciousness and rivalry. As regions are involved more closely with each other, differences are perhaps more apparent.

It should be made clear, however, that the concept of the region is primarily an intellectual device, useful as a tool for the better understanding and analysis of human society. The regional concept picks out certain spatial patterns, but even political boundaries or other regional distinctions we may draw are products of our own minds rather than complete descriptions of reality, or of any "natural" order. In this sense, there are no "natural" regions, but only man-made or man-perceived classifications of the welter of areal differences.

The regional hierarchy

The differentiation of regions may begin very low on the size scale, and it is impossible to establish the order of magnitude in which one is dealing in a particular case. Differences which one may call regional exist, for instance, between different parts of a city, and at the other end of the scale they are apparent between continents and hemispheres. Two neighborhoods or two cities will, however, have much in common as part of a larger urban, provincial, national, or even continental region. There is thus an overlapping hierarchy of regions, depending on the criteria used. One may look at the world or at any part of it through the big end of the telescope to see gross general spatial patterns, or through the small end to see more detail and more minute subdivisions. Both devices are useful. But the fact of areal difference is universal, and is expressed in ramified details of culture, from diet to dress, house types, language, prejudices, tastes, occupations, skills, and social and political systems. The United States is composed not only of a hierarchy of political subregions in the form of wards, cities, counties, and states, but of a great variety of more broadly based subregions involving areal differences in land use, employment, investment, production, voting habits, income, marketing patterns, cultural attitudes, speech, and a host of other characteristics which have spatial form and are discernible as spatial patterns. Some of these broader spatial patterns are commonly recognized in geographical expressions as "the South" or "New England," but each of these regions may be further subdivided not simply on the basis of state lines but in recognition of more complex variations from one part of the South or New England to another. At the other end of the size scale, it is clear that both of these American regions as a whole belong within the larger region of the United States and in turn of North America; at that level, or on a world scale, their similarities become more important than their differences.

Bases of distinctiveness

Regional differences may in part be related to differences in the physical environment, but only in part. Human society is a complex affair in which a great variety of factors operate. The regional analyst must attempt to consider all of these factors. An area covered by a given climate or soil may support a more or less consistent type of land use or culture, related in part to climate and soil, but related also to non-environmental factors and indeed often occurring in physically different areas. A region may derive much of its distinctiveness from its interconnections with other areas. The Netherlands, for example, is a distinctive region in part because of its widespread trade relations with the rest of the world. The Rhine River, which is partly responsible for the commercial prominence of the Netherlands as a break-in-bulk point at the river's mouth, is also responsible for considerable economic unity in the Rhine Valley. Although political boundaries cross the Rhine and in some stretches use it as a political line, a heavy flow of trade follows the river. It is possible to argue that the "Dutch" region or the "Rhine" region actually extends beyond the limits of the conventional political or physical lines and includes aspects of the area with which the region trades. This kind of analysis, however, brings us back to the essential problem that few aspects of a region are likely to co-vary exactly, or to cover the same area, and that the application of a regional label to any area is ambiguous until the criteria used are first made clear.

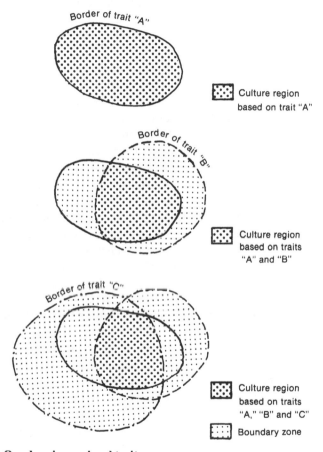

Figure 1 Overlapping regional traits

The example of China

Take, for example, the case of China. China as a political state covers a vast territory, larger than the United States and containing within it a great variety of physical and human differences. About half of this area is occupied predominantly by people who are ethnically not Chinese, speak a non-Chinese language, and have a non-Chinese culture and economy. Over half of it is so dry, so cold, or so mountainous that it contrasts sharply with the relatively well-watered, warm plains where most of the Chinese people live. On the other hand, prominent aspects of Chinese culture and large blocks of Chinese people are found in several places in east Asia beyond China's political frontiers: Chinese written language and culture forms in Japan, Indo-China, and Korea, heavy Chinese settlement in Thailand, Malaya, and Java. Where is the true China region? The answer depends on the criteria chosen, or on those to which the greatest weight is given. Generalized lines may be drawn on the basis of single factors – around the area where Chinese-speaking people are in the majority, or where agriculture is the predominant occupation, or where the rainfall exceeds twenty inches a year. This may be a useful service, but it begs the major question. What makes China China, and where does it cease to be China?

One convenient solution is to plot on a map, as in Figure 1, the areal distribution of all factors which could be considered relevant as measures of regional identity: consistent language, physical type, economy, climate, land forms, social and political institutions, political control, cultural forms, historical alignments and changes, and so on down the list of any region's characteristics, although these will be of different relative importance for different regions. The basic minimum area to be included as indisputably China would be the area where all of the factors designated as consistent Chinese core characteristics were found. But none of these characteristics covers exactly the same total area, and somehow the margins must be disposed of. A decision must be made about which of the characteristics are most important. Areas where they are all absent can be eliminated; areas where only some of them are absent will have to be treated on the basis of the priority given to each characteristic as a measure of Chineseness. By such means one can ultimately arrive at a reasonable regional differentiation, but only after it has first been decided what makes the region distinct. The plotting of areas on a map as suggested above will usually help to suggest which are the critical factors, but it will also be necessary to make a thorough analysis of the regional *Gestalt*, in order to avoid an arbitrary decision and to understand adequately the proper relative importance of the factors which must be evaluated. This is obviously not a simple problem, for all the ease with which lines are often drawn or the glibness with which people talk about "natural" regions. In the case of China there is a strong lead because Chinese civilization has traditionally rested on agriculture. Where agriculture and the farmer stop on the map, there "China" has tended to weaken or disappear. In this case most of the essential regional characteristics co-vary with agriculture in their areal distribution. In the excessively dry, cold, or mountainous parts of political China, agriculture is greatly reduced or eliminated, and many of the other "China" characteristics are replaced by a clearly non-Chinese set of characteristics, although since 1950 the Communist government has fostered much cultural uniformity.

It is then logical to ask why Chinese political control extended so far beyond the limits reached by the other factors, and why political control was not extended over the areas in other parts of east Asia where Chinese people or Chinese culture did spread. This question is raised not in order to answer it, for that would be beyond the scope of this brief example, but to emphasize the implications of the regional concept and the insight which it may stimulate.

Regional cores

Associated with the fact that in most regions the essential characteristics gradually shade off as the regional margins are approached is the further common feature of regional cores or nuclei. Usually spatial patterns of human development tend to focus or to coalesce on a hub or axis. This may be a physical phenomenon such as a fertile plain, a river valley, a sea-coast, or a crossroads of easy routes. Or it may be a cultural development such as a city, the center of a particular type of regionally dominant economy, or a center of political or military control. Most clearly marked regional cores have all of these characteristics. They coincide with the largest area of dense population, include the largest share of the region's economic activity and the political capital of the region, and tend to set the region's cultural pattern. Usually this core area is the first in point of time to develop the cultural characteristics of the region as a whole and it spreads its dominant influence over the surrounding area as far as the original physical and cultural conditions allow. At some point these conditions become different enough or far enough away so that they cannot be adapted by influences exerted from the core. This distant area will lie in a zone of transition or will be attracted and influenced more powerfully by another less distant and more congenial center and will belong to that center's region. Most regions seem to be given much of their coherence by comparatively well-defined cores around which their areal traits tend to cluster.

Prominent examples of regional cores include the London Basin and the Paris Basin. Both are the chief centers of population in their respective national regions. Both are the centers of political control and the major market centers. Both acted historically as nuclei for the development and spread of their respective regional cultures, and both are still the dominant nuclei in this respect. Transport lines and circulation patterns focus on both to a high degree. The pattern of circulation and the density and degree of interaction within the region may often be a leading part of or clue to regional identity and distinction. Regional lines may be drawn on the basis of orientation in circulation patterns to a particular regional core, or on the basis of common interaction. Beyond such a line, orientation may be predominantly to the core of another region, even though it may be distant, and to a different set of interaction partners, although there may be a zone of ambivalence or indifference lying between competing cores. London and Paris are both strong cores whose dominance extends throughout and to some extent beyond their respective national areas. But in both countries there are also subsidiary cores, most importantly the centers of heavy manufacturing which developed much later than London and Paris and which are areally separate, associated with mineral resources largely lacking in the London and Paris basins. These lesser centers are subsidiary or tributary to London or Paris in part because they are mainly single-factor cores rather than multiple or generic in their nature. In other regions or countries more evenly balanced competing cores may develop as, for example, in Germany (Berlin and the Ruhr), and to some degree in the United States (the middle Atlantic coast and the Great Lakes-Chicago district). As with the hierarchy of regions, however, there is a hierarchy of cores which may be distinguished at different orders of magnitude. Strictly speaking, a core performs some service or exerts some influence over the whole of the larger region to which it belongs. Manchester and the English Midlands are subsidiary to London, and Chicago and the Great Lakes district to New York, but these subsidiary cores do play a prominent national role in the provision of certain major goods and services. St Louis, however, or Atlanta, although they are centers of settlement cores within small subregions of the United States, do not play a marked national role and hence are not cores in the larger regional sense; they are merely centers of population clusters and of certain local functions or local regional characteristics.

Space, time, and change

The delineation and analysis of regions is undertaken by geographers because it involves the distribution and pattern of things on the earth's surface. They are concerned with matters whose occurrence can be mapped. But they are also concerned with the relationships between regions, with the association of areas, and with spatial interaction. The patterns which they map are not static, nor are they self-sufficient. They are part of an integrated whole which it is just as important to fit together as it is to take apart. Before one can examine relationships within the whole, it is necessary to distinguish the parts. This involves not only separate forces and factors but the concerted interplay of the whole complex of factors which makes up each region and the areal associations which result.

In drawing regional boundaries, analyzing regional units, and considering the association of areas, the time dimension is of critical importance. Regional boundaries are seldom constant, since they reflect a particular stage of development. As the aspects of a society which give it its distinctiveness change, the degree and nature of its spatial expression change also. This is the result not only of technological change, such as improvements in communication which can extend unity, but of other temporal changes in the basic nature of the community. Wholly new factors or new emphases may arise in the economic or political system, for example, which produce a different core, the exclusion of former areas, or the inclusion of new ones. Invasions, conquests, or migrations may distort or basically alter regions. Cultures change, are powerfully influenced by other areas, or may be imported. New forms of production may arise to dominate the economy or to revolutionize its nature. These kinds of changes in human settlement and use of an area, including changes in the nature or identity of the settlers, are called *sequent occupance*.

There tends, however, to be a certain generic harmony of the components of any region which may reassert itself, although perhaps in a somewhat different form, even as the components change. Only over relatively long periods or under special circumstances do regions vanish or wholly new ones arise. The physical environment is relatively constant, and even though its influences change as people change, it helps to maintain a certain amount of regional continuity over time. Spatial relations are also manipulated and altered as people change, but physical distances and absolute location remain constant. While the "China" region has remained relatively constant areally during the past two thousand years (despite recent expansion into Manchuria and Inner Mongolia), tremendous changes or outside influences in Europe and North America have destroyed most of the regions of two thousand years ago and have produced new regions in their places. The same temporal processes are at work, slowly or rapidly, in any current situation, continually changing the essential quality of every region and thus affecting the region's spatial limits.

Single- and multiple-factor regions

Although we may speak in general terms of China or Mexico or Canada as regions, such a label must of course grossly oversimplify the complex of culture and environment involved in such large areas. We can more accurately speak of *single-factor regions*, such as political, linguistic, or climatic areal units, or we can outline *functional regions*, such as trade or administrative areal units. The United States is divided, for example, by the Federal Reserve Bank into twelve districts, each intended to include a recognizable and viable commercial-financial region grouped around one dominant urban financial center (see Map 1). In another pattern of *functional regions*, the Sears Roebuck Company divides the country into five major and eleven minor merchandising districts, each with its own urban center for warehousing and service for the district as a whole; each of these districts offers a slightly different set of goods, reflecting differences in regional needs or tastes and consumer preferences. The division of the United States and Canada into states and provinces is an obvious example of political administrative regions. Broad patterns of climatic differences are recognized in climatic regions. Single-factor regions such as these can of course be sharply delimited with a single line. A political boundary or the boundaries of a Federal Reserve District or of a precisely classified type of climate can easily be defined and will outline simply and clearly *one* consistent spatial pattern, extracted from the complex matrix

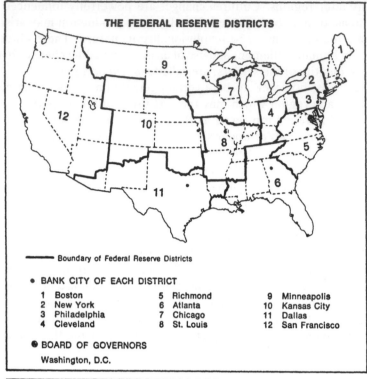

THE FEDERAL RESERVE DISTRICTS

———— Boundary of Federal Reserve Districts

• BANK CITY OF EACH DISTRICT

1	Boston	5	Richmond	9	Minneapolis
2	New York	6	Atlanta	10	Kansas City
3	Philadelphia	7	Chicago	11	Dallas
4	Cleveland	8	St. Louis	12	San Francisco

⊙ BOARD OF GOVERNORS
Washington, D.C.

SEARS ADMINISTRATIVE DIVISIONS

———— Boundary of Administrative Divisions
⊙ Division Headquarters

Map 1 Sample regionalizations of the United States

of reality. But there is an undoubted unity implied and apparent in broader geographic expressions like "the South," "New England," "the Sun Belt," or "the Great Plains" applied to very much more varied constructs including a great number of different characteristics. It may be difficult to define or delimit this unity, aside from obvious political boundaries. Each of the factors involved will almost certainly cover a different area, overlapping each other near the center but leaving a good deal of the "region" ambiguous. The American South, or New England, for example, are all *multiple-factor* regions whose identity depends on a variety of individual characteristics more strongly and unanimously present in some parts (Alabama, Vermont) than in others (Kentucky, Connecticut) which may lie spatially close to the edge of the region or may be influenced by adjacent regions. But the idea of the region assumes the existence of some spatial unit which exhibits a set of common traits. This broader concept of the region might be stated as a *definable area possessing a wide range of more or less homogeneously distributed and interrelated characteristics*. It is seen as a complex of homogeneous factors which are interdependent and which operate together as part of a larger whole, or *Gestalt*. The idea of the region has often been criticized because

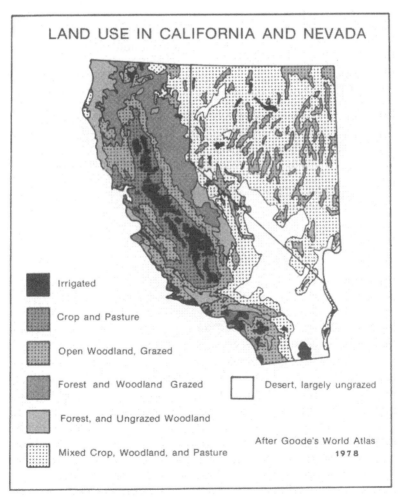

LAND USE IN CALIFORNIA AND NEVADA

Irrigated

Crop and Pasture

Open Woodland, Grazed

Forest and Woodland Grazed

Desert, largely ungrazed

Forest, and Ungrazed Woodland

Mixed Crop, Woodland, and Pasture

After Goode's World Atlas
1978

Map 2 Land use in California and Nevada

it is ambiguous or oversimplified. The distinction between single-factor or functional regions on the one hand, and broader, multi-factor generic regions on the other hand, helps to avoid much of this criticism. The general concept of distinct areal homogeneity is extremely useful, provided it is based on an analysis of the many factors which make an area distinct.

Planning for the developing of any area must depend on detailed knowledge of what functions, resources, population, and other characteristics the area includes, plus the nature of its relations with other areas. Comprehensive inventories and analyses of all these factors should precede any effort to introduce changes. Regional planning tries to do this and to produce, for example, detailed studies of the environment, of existing land use (see Map 2 and Figure 5), or of trade flows. Data such as these are then used as the matrix into which new developments, such as a manufacturing plant, a transport facility, an irrigation project or new crop complex, a recreation area, or a hospital can be introduced with maximum benefit. Such decisions must also rest on knowledge of all the area's characteristics, including its environmental base, transport networks, the spatial distribution of its population, its income levels, employment patterns, and so on. Urban planning is similar on a micro scale, and is concerned among other things with insuring the orderly growth of cities so as to provide essential services in a well located pattern and to minimize the juxtaposition of incompatible activities, such as heavy manufacturing and residence. Planning tries to ensure an appropriate mix of employment, housing, retailing, public transport, park areas, schools, and so on. *Zoning laws* are designed by urban planners to designate specific urban areas for specific uses and to try to prevent incompatible mixes.

Study of regional wholes

Regional information and understanding are objectives of the geographer as well as the classification of regions. A large part of geographic work consists of the accumulation of regional material. Here also geography undertakes a task which is generally not done *holistically (as an integrated whole)* by other disciplines. The economist is interested in production, finance, or the market system in any region, the sociologist in kinship groups, social structures or institutions, and so on. Geographers attempt to examine and to understand the region as a whole, in all of its physical and human aspects, to fit these aspects together, and to relate the region to other regions. This is a demanding goal and may never be completely met, but someone must make the attempt to fit together the regional mosaic and to see regional wholes. So much has to be examined and understood in order to do this adequately that geographers must depend on the specialized studies of others as well as on their own studies, and they frequently function to some extent as editors, or integrators.

They must also compile and analyze a great bulk of regional information. This collecting and editing is an essential but elementary first step, a means to an end. Other students of society need the information which geographers can provide. Regional planners, the general public, and those who make policy decisions of many kinds need regional information too. But the information alone is not fully useful until it can be fitted into the regional whole. It can also be dull to deal only with factual detail, to see only the individual trees and miss the pattern of the forest. Sometimes the "pack-rat mind," which collects endless bits of information for its own sake, is apparent in the craze for facts and in the confusion of information with knowledge which handicaps some regional study. Information can seldom be gathered intelligently unless an over-all pattern is seen. Buildings are constructed of bricks, but bricklayers cannot, or should not, choose and lay bricks without a blueprint or an image of the building. Knowing the number of miles of railroads in Chile, for example, is less important than knowing the patterns of distribution and interrelation which the railroads reflect and stimulate in Chile, the role of railroads in the regional economy, and the part which these patterns or factors may play within the whole of the Chilean regional unit. But the simple information is necessary first, and the need for that particular information is fully apparent only when the larger regional context is understood.

Holistic regional analysis attempts to delineate the particular use which people have made of the earth in a given region and the pattern and effect of spatial relations within it and between it and the rest of the world. Such analysis usually centers on population and its distribution, density, and growth, the nature of the settlements, the agricultural system, the use of other resources, the pattern and nature of manufacturing, trade, and transportation or circulation, and the nature and distribution of the regional culture as a whole. Consideration must also be given to historical factors and changes and to everything of significance about the region which has spatial form or spatial distribution. Regional analysts must in many cases depend on the work of particular specialists. Climate will be important to them, for example, but they will probably turn to climatological work for the detailed facts and processes, as they will in other cases turn to the work of soils scientists, agricultural economists, anthropologists, or historians. They must themselves fit these varied data together within a regional framework which it is their business to provide. The regional whole must be understood as the context in which any of its parts are to be examined, since all of its parts are interrelated. A specialist's approach to agriculture, for example, might make much sense of the Russian agricultural system which a more general regional approach might miss; but it might also make much nonsense because the specialist did not adequately evaluate the influence of politics, tradition, or social organization on agriculture in this particular regional situation. Each region must be treated as a single areal manifestation of an integrated body of culture.

This kind of regional understanding is the basis for regional comparisons, which may throw light on whatever is being examined, whether it is a single aspect of the region or the whole complex. Comparison is obviously necessary in making regional distinctions. Historical comparison is necessary, too, for any region is a dynamic entity whose nature is continually changing.

One cannot speak geographically without speaking of places, and one cannot speak effectively of places without considering the regional concept and the pattern of areal relationships. Special analysis of industrial resources, transportation, agricultural types, settlements, or political factors is not only incomplete but liable to serious error until it can be applied within a variety of regional frameworks and regional contexts. This means more than superimposing data or ideas on a landscape, or extracting them from it; the regional whole must be understood. Regional models must be constructed against which any assertion or study or discovery with regional relevance can be measured.

The progress of knowledge is not served by insulating the boundaries of convenience which separate its various branches. No student of society is self-sufficient, and all of the branches of study are interdependent. Within geography, any sound analysis must make thorough use of both the topical and the regional approaches as well as depending on contributions from other disciplines. But it is the nature of geography that at the heart of all its work is the concept of the region.

3 SPATIAL FORM AND SPATIAL INTERACTION

Geography asks how and why people and their activities are distributed as they are: what accounts for the great clusters of population and for the empty areas, for the cleared and uncleared land, the cities and other settlements, the industrial areas, the trade routes. It looks for spatial patterns and studies the interactions there may be between them.

If the roughly five billion (US) people of the world in 1982 were distributed evenly over the earth's land surface, there would be an average population density of about seventy-six people per square mile (33 per square kilometer). If they were gathered together as closely as an audience in a theater, 300 square miles (550 square kilometers, or about a third of the state of Rhode Island) would hold them all. As it is, about half of the earth's land area supports fewer than two people per square mile, some densely populated rural areas support as many as three thousand per square mile (1100 per square kilometer) and population density in the centers of large cities may reach or exceed a hundred thousand per square mile (38,000 per square kilometer). Settlements of various kinds are scattered widely over the earth, but only a few have become big cities. In some rural areas of the world, farmers live in villages *(nucleated settlements)*, and in others they live in isolated farmhouses *(dispersed settlements)*. The tea shrub, like many other commercially useful plants, will grow in many different places, but only in a few areas is tea actually grown on a large-scale commercial basis. Mineral ores and fuels occur in scattered locations over much of the earth, but heavy metal manufacturing has developed on a large scale in only a few places. There are approximately three thousand different languages in the world; some are spoken only by a few people, some by hundreds of millions, some are limited to a single large or small area, and others have spread over widely separated areas. What are the implications, causes, and interrelations of these varied spatial patterns?

Geography is concerned wherever the fact or the concept of *place* or the question *where?* is involved. Places and spatial patterns, as in the examples cited above, are not interpreted by geographers as static phenomena or seen as isolated facts, but in terms of relationships. It is only of incidental importance where Timbuktu is (or China, or the wet tropics, or the area where Swahili is spoken), except as a knowledge of the location helps to understand the relation between Timbuktu and the rest of the world and the effects which these relations have had on Timbuktu and vice versa. No place and no pattern of distribution in the world is exactly like another place or pattern, but there are marked similarities among groups of places and spatial patterns. In each case, the nature of relations with other places may be just as important as local conditions in accounting for the differences or the similarities. Geographers attempt to analyze human society in its spatial frame. They examine not only where things are and what forms or groups may be discerned but why they are there and how their location and their nature and development may be related to the influences exerted by other places. This is the meaning of the term *spatial interaction*.

The development of theory which can help to explain real spatial patterns can often be aided by imagining the earth as a smooth, featureless ball on which people and resources are evenly distributed. If the varied elements of complex reality are added one at a time to this simplified image, the consequences of each element may be deduced, and some of the aspects of experimental science simulated. Location theory, in particular, makes use of this technique as in the development of the theory of the *central place* discussed below. The understanding of complex reality requires the kind of generalization which theory makes possible and out of which theory is constructed. But generalization, and the oversimplification on which it necessarily rests, must continually be tested against specific pieces of reality.

Spatial relations

Timbuktu

Timbuktu is a real city in Mali in northern Africa (see Map 3). Why is it where it is, and why has the settlement on this site grown big? The reason for its location close to the Niger River is clear. Timbuktu lies near the margin of the Sahara Desert, and, in the absence of adequate rainfall or lakes and of ground water for wells, settlement in this area must be limited to sites where water is available. *Site* refers to the actual ground on which a city or some other phenomenon rests. It is often paired or contrasted with *situation*, which refers to spatial relations with other places, principally in terms of accessibility rather than of simple distance. Situation is thus both a relative and a changeable matter. Timbuktu's situation – its relations with other places – has been much more important than the local conditions of its site in accounting for the city's growth and

Map 3 Timbuktu in West Africa

decline. Northward is the desert. Southward the land becomes more humid and therefore more densely populated, with a type of productive economy different from that of the desert. Settlements which lie in the *zone of transition* between one set of physical or economic conditions and another set may be convenient centers for the exchange of the different sorts of goods which each area produces. In part, Timbuktu has prospered as a trade center for this reason. The city also has the advantage of easy access to the moister and more productive area to the south by means of the Niger River, which flows southward to the coast. Towns and cities are specialized, concentrated settlements which have grown primarily because they have access to wider areas around them, referred to as their *hinterlands*. They thus can perform on behalf of these wider areas economic, political, or social functions which are most conveniently carried on in a central place, such as government, marketing, manufacturing, or the servicing of trade routes. The differences between the physical natures of the areas to the north and south served by Timbuktu promoted a *complementary* trade relationship between them, since they produced and needed to exchange different kinds of things. Regional differences do not always result in exchange. The amount of exchange depends on the effective demand in one region for the commodities produced for sale in another region and on the cost of transport between the two places.

Timbuktu is a convenient center for the complementary exchange between the different areas around it. This exchange is due not only to its median location, but because transport means are likely to change there, from land to river, from desert transport to humid-area transport. Timbuktu has in fact been called "the place where camel meets canoe." Places where the method of carrying changes (or *break-in-bulk points*) will give economic opportunities to their populations. This will apply not only to loading and unloading, storage, or servicing of transport carriers but also to the financing and insuring of trade and to the processing or manufacturing of goods passing through. Goods must be unloaded at the break-in-bulk point; transport and loading-unloading costs for goods are therefore minimized at that point, and a variety of routes are likely to focus there, funneling goods through the city. These are important matters for any manufacturing or service industry dependent on the assembly of raw materials or the distribution of finished goods. The trade through Timbuktu has never been very great, by modern standards, because the tributary areas (hinterland) which the city serves are sparsely populated and unproductive (the desert), or largely organized economically on a subsistence rather than a commercial exchange basis (the moister areas to the south). Hence Timbuktu is not a very big city, but it is the biggest and almost the only city in this part of northwest Africa.

Timbuktu would be very much smaller today, however, if there was only desert to the north. The desert produced almost nothing for exchange, but beyond it is the moister Mediterranean coast with different produce from the areas south of Timbuktu. This coastal region exchanged goods with humid central Africa over the most practicable route, which goes through Timbuktu. Some of these goods were redistributed from north Africa by cheap sea transportation to wider markets in Europe and elsewhere, where under different physical and cultural conditions the same goods, such as ivory, gold, spices, or slaves, were not produced. These wider relations were reflected in the size and functions of Timbuktu, hundreds or even thousands of miles away. The goods which moved along the route north of Timbuktu had, however, to be of a certain sort. Here there was no river to lessen transport costs, and it was a very long haul to the Mediterranean coast, through an area largely empty of population and devoid of opportunities for exchange. The carriers (camels) are very expensive, measured by the unit cost of transport, or the cost per ton-mile (for carrying one ton one mile), mainly because their individual carrying capacities are small. Transport costs tend to vary inversely with the capacity of the carrier, which is the principal reason why water transport is so cheap. The length of the route between Timbuktu and the Mediterranean coast increased the expense, as did the lack of trade-generating centers along the way. The latter are important factors in making a transport route profitable. Therefore this route could carry only goods which were very high in value by weight, such as gold, ivory, slaves (who provided their own transport), or salt (which is very high in value in most preindustrial economies).

With the development of mechanized sea transport and the widening of Africa's commercial relations with the rest of the world, the goods which used to move through Timbuktu to wider markets began to move instead through west African seaports. Timbuktu has become a much smaller place and more exclusively dependent on local trade and services. Change, especially technological change, is continually altering spatial relations and spatial interaction patterns. At its zenith about the fifteenth century, Timbuktu was a commercial metropolis. Because of its size and its excellent access, it was also a famous cultural center for much of the large area of the western African Sudan, and supported a well-known Moslem university. With the deflection of the trade routes which fed it and their interruption by political disorders, Timbuktu's glories faded. By the end of the nineteenth century, the French found it a largely ruined town. Both political and technical change were thus responsible for Timbuktu's decline, as they affected the city's relations with other places or its pattern of spatial interaction.

Other examples of spatial interaction

California's agriculture is now specialized in the growing of fruit and vegetables partly because of a favorable climate, but also because in the course of the last century California has developed cheap, easy, and quick access to a large market in the eastern United States, where the climate is different. Without the east, if it were a different and less complementary kind of place, or if it were not for the railroads, California agriculture would be a smaller and very different enterprise. The effects of weak spatial relations may be equally important and obvious in other situations. Isolation was as responsible as environment for the uniquely primitive level of the Australian aborigines and for the low level of technical development characteristic of the North American Indians and of the various groups native to the rain forests of Africa and South America.

Analysis of the relations between places can be used to throw light on every problem which the geographer investigates. It is part of geography's principal function, the study of space and place. Comparison is also an essential part of geographic analysis and interpretation. The two great poles of world civilization, Europe and the Orient, cannot be fully understood until they have been compared. Comparison helps to reveal the distinctive qualities of each and may suggest some of the distinctive dynamics which, without the comparison, may be taken for granted or overlooked because the areas have been examined only in their own respective contexts. One's own region or country can even less well be understood and interpreted until it is compared with some knowledge of other regions or countries and thus seen in some perspective. A knowledge of only one area or society may also be seriously misleading if it is applied to the interpretation or evaluation of another area or society. Use of only a single comparative criterion without knowledge of the wide differences elsewhere (and without realizing that there is no single cultural norm) perverts the purposes and benefits of comparison, and it is regrettably common among Americans, whose knowledge of most of the rest of the world is dangerously limited. The comparative method is of course no special property of geography. Anthropologists and sociologists in particular use it as the basis of much of their work, and it is as important in history as in geography. The geographer merely applies the comparative method to the problems of spatial form, spatial relations, and regional differences.

Geography and history

The classifying, describing, and interpreting of spatial patterns which geography attempts, including the division of the earth into regions, is the principal basis of the frequently made analogy with history, which classifies, describes, and interprets patterns in time. Every investigation of human society is more useful if it can place its particular inquiry within the double frame of historical time and geographical space. Neither historical nor geographical analysis is complete unless it critically applies *both* dimensions. Historical changes alter spatial patterns and spatial relations. No landscape and no pattern of land use is static; they are continually changing, and show varied and cumulative effects of historical developments in the recent and remote past. A geographic analysis which ignores such factors, or which attempts to treat a landscape or a spatial pattern as if it were static or free from change in the future is a travesty of reality. The examination of such changes, their rates, directions, causes, effects, interrelations, and whether or not they are cyclical or recurrent, is as much the geographer's job as the historian's. Correspondingly, spatial and environmental factors affect historical changes. Historical analysis which ignores such factors gives an incomplete, distorted, or at worst lifeless and meaningless picture, since it neglects the stage on which history takes place. "Geographie without Historie hath life and motion, but at randome and unstable. Historie without Geographie, like a dead carkasse, hath neither life nor motion at all."[1]

Cities and spatial interaction

Let us return to the case of the city, perhaps the simplest and clearest example of spatial form and spatial interaction. One may take as given the facts of its existence, its population, number of houses and factories, miles of paved streets, and the size of its waterworks. These can be found in a guidebook. But to explain these facts and to explain the spatial form of the city, one turns first to its relation with other places. The city is

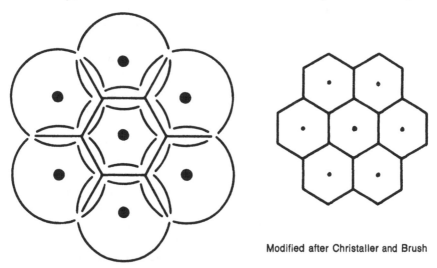

Modified after Christaller and Brush

Figure 2 Hypothetical spatial patterns of city hinterlands

as big as it is and specialized in its particular way because it provides services for a particular area dependent on it, the city's *hinterland*. The size of this tributary hinterland is determined by the ease of access to the city, plus the competition of other cities which might also serve the same area (see Figures 2 and 3). The total productivity of the hinterland and the size of its market are matters over which the city can exercise little influence, except as it can improve its own access. It is the hinterland which determines the city's size and specializations. What grows up in one urban center is primarily a reflection of access, of spatial factors over a very wide area, and of the interrelations which take place within it.

The size and productivity of the American Middle West and the easy routes which lead from it through the Mohawk-Hudson corridor are in large part responsible for the size and prosperity of New York City. Provincial residents of Manhattan may find it hard to accept the Middle West as a major factor in their existence, but this shortsightedness is characteristic of most city dwellers. In most big cities the butchers, bakers, and candlestick makers for the city's own population employ more people than are employed in the so-called *basic* enterprises, which are producing goods or services to be sold outside. It is these basic enterprises which support the city and make it necessary to have butchers and bakers, cab drivers and street cleaners. It has been estimated that one person employed in a basic enterprise, such as wholesaling, manufacturing, or insurance for areas outside the city, will support approximately two other people employed in the manufacturing and service enterprises supplying him and his family with what they need: food, schooling, department stores, housing, and so on. This roughly one-to-two relationship between employment in production for external sale and employment in production for internal consumption is called the "basic-service ratio." But we cannot all live by taking in each other's washing, even though most city people may do so, and the concept of basic enterprises reemphasizes the fact that cities exist in order to perform certain specialized functions on behalf of a wider area. A big city like New York will in fact overlap the hinterlands of other cities because it usually has highly specialized goods or services to offer which can be obtained nowhere else, while for less specialized things a smaller and nearer city will be the center. Specialization of any kind, even in a village general store, may reflect particular local advantages or may be simply the result of bigness, but neither of these can be implemented unless the village or city has effective connections with a hinterland tributary to it.

Various attempts have been made to explain the distribution, location, and size of cities and towns, working from the assumption that urban settlements arise to serve a more or less well-defined hinterland. Attempts have also been made to discern consistent spatial patterns of city location and size, on the assumption that there may be a regular order of distribution. From some of these investigations there has been developed the theory of the *central place*, which argues in part that cities may be spaced at more or less regular intervals, each serving a hinterland around it whose limits are fixed primarily by distance. Where distance is the only factor, hinterlands would theoretically assume the shape of a circle whose center is occupied by a city – or since several competing cities actually occupy the landscape, the hinterlands would be tangent on all sides and would thus have the shape of regular polygons, as shown in Figure 2. But distance alone is rarely, if ever, the only factor affecting access, and in reality city hinterlands are grossly irregular in both size and shape, the cities themselves differ widely in size, and the pattern of city

Figure 3 City hinterlands in a hypothetical landscape

distribution is far from regular, as the physical and human landscapes which they serve are not regular or uniform. Mountains, rivers, plains, seas, deserts, and other features of the physical landscape affect access or spatial interaction. Human factors, differing widely from place to place, and especially technological factors as they may influence transportation and production, also affect access. Differing degrees of complementarity between different regions, on both physical and human grounds, are further reflected in differing degrees of spatial interaction. Political boundaries, cultural differences, energy costs, or linguistic barriers, among many other matters, affect the flow of goods, services, and ideas, and thus also affect the distribution of cities. Among regions, productivity differs, demand levels differ, and each region tends to require a different arrangement and different sizes of cities to serve it. A large semi-arid area with a sparse population and low productivity, or a large area occupied by a dense but mainly subsistence or self-sufficient population, cannot support as many or as large cities as a smaller but more productive and more commercialized area.

The theory of regularly spaced cities and hinterlands is also disrupted in reality by the clearly observable *hierarchy* among urban places. A large city with good long-range access, New York, for example, may include in its hinterland most of the other cities in the country, to differing degrees. For large capital loans, exclusive dress fashions, or specialized publishing, New York is in many cases the only supplier, and for these things even Chicago is tributary to New York, as most of the Middle West and the larger area served by Sears Roebuck, for example, is tributary to Chicago. Cities of middle rank may similarly include other smaller cities in their hinterlands. At the bottom of the size scale, a rancher in Wyoming, for example, will obtain the bulk of his goods and services from the nearest or most accessible small town in whose immediate hinterland he lives, but he will also be tributary to Cheyenne (the state capital), to New York for cut diamonds or books, and to Chicago for barbed wire and sheep or cattle brokerage. If his wife wants a new dress, she will probably shop for it not in the small town, but the nearest or most accessible small city, whose hinterland for more specialized goods and services overlaps and incorporates that of the small town.

When a city grows big because it serves a productive hinterland, it acquires more and more specialized functions which then further increase the city's size as they enlarge its hinterland. The biggest cities thus tend to have a self-perpetuating advantage. But for any village, town, or city, tributary hinterlands tend to be different for each type of commodity sold or service performed. Medical services will cover one area, food retailing another, wholesaling another, and sale of heavy manufactured goods another. Where manufacturing becomes highly specialized, even a relatively small city may sell its products over a very wide area. For example, Hershey, Pennsylvania, and Hollywood, California, include large parts of the world in their respective hinterlands in this specialized sense, and the same may be said even of the tiny island of Harris off the coast of

Scotland, which is world famous for tweed, even though it is poor and sparsely populated. In all of these cases, urban location and urban growth reflect spatial interaction. Relations between places are affected not only by distance but by differing degrees of access, by the complementarity which results from regional differences of all kinds, and by the simple factor of the size and consequent specialization of urban places, which is in turn a reflection of access.

An oversimplified example may help to illustrate the idea of access and the multiplicity of factors which affect patterns of spatial interaction. In Figure 4 a man living at A may choose between the cities at B and C to

Figure 4 Simple interaction diagram

obtain the goods and services he requires. B is closer in physical distance, but a mountain range intervenes. The shortest way there in distance is by road, but the cheapest route for bulk shipments is by the railroad over a longer route. If the man has a car and is in a hurry, he will travel to B by road, but most of the bulky goods he orders from B will travel via the railroad. There may be other factors, such as the location of intermediate services or shopping places, the enjoyment of scenery, or social and personal benefits which make one route preferable to the other. C is a larger city, however, in part because it lies on a productive plain, whereas B's hinterland is restricted in size by mountains and is not as productive per square mile. The man at A will therefore find a greater variety of goods and services at C, and despite its greater distance will probably obtain there most of what he needs. In reality, the situation would probably be much more complex. Political boundaries, language barriers, or artificial rate structures might intervene which would distort flows between A and C. Cultural differences between A, B, and C, special local skills, traditional alignments, or differences in law might influence flows in a particular direction. The man at A might also quite likely split his spatial relations between B and C, being tributary to B for certain goods and services and to C for others. But even in the limited case of the example used, time, distance, costs per mile, intermediate opportunities and rewards along the various routes, the nature of the local and regional environments, and the size and nature of the respective cities, plus other less well-known and more varying factors, all contribute to the pattern of spatial interaction which takes place.

Spatial interaction is by no means confined to cities. Wheat is grown in the Dakotas and forests are worked for lumber in Oregon because of the relations between these places and the rest of the world. A century ago wheat growing and lumbering were absent or minor in these places, although the climate, soil, and relief were essentially the same then as now. What has changed is that the Dakotas and Oregon have been connected by transport networks to highly developed distant markets, especially in the eastern United States and northwest Europe. These markets demand wheat and timber and have the economic and technical means of satisfying their demand from distant sources. They no longer grow most of their wheat and timber locally, although physical conditions there may be as favorable or more favorable for these crops than in the Dakotas or Oregon. Other activities are now more profitable for Pennsylvania or England as a result of the industrial revolution (see Chapter 15 for a discussion of this term) and of the wider relations between Pennsylvania and England and the rest of the world. Wheat or timber may be obtained from distant sources more cheaply, or local land, labor, and capital may be freed for more productive uses. Wheat yields very well in the climate of northwest Europe or the eastern United States, but wheat is tolerant of climates which are too dry for many crops, and hence it is often pushed out toward the dry margins, leaving the moister climates for more fussy and more profitable crops. But the spatial forms which have grown up at both ends of the line, agricultural, manufacturing, commercial, or others, are in large part the result of spatial interaction. What each area is doing depends on what a variety of other areas are doing and on the low-cost transport which ties them together.

Change and spatial relations

One final example of the importance of spatial relations may be cited. For the first two thousand years, more or less, of Western history, the focus of development was the Mediterranean Basin. The Mediterranean is a virtually enclosed sea, relatively free from storms, easy to navigate because one is not out of sight of land for very long, and yet connecting a great variety of diverse regions. Three continental culture areas come together there, so that after people learned to sail its quiet waters, the Mediterranean became an easy avenue of exchange very early in Western history. Exchange is necessary for economic specialization and is thus a condition of economic development. Toward the end of the European

Middle Ages, however, new techniques of navigation arose. The Age of Discoveries revealed in time a world larger than that the ancients knew, with its focus in the Atlantic. The most accessible place was no longer the Mediterranean but the shores of northwest Europe, from which multiple sea routes led to vast and productive new worlds toward both the east and the west. It was this change in spatial relations, rather than the much later industrial revolution, which was mainly responsible for the areal shift of human emphasis which took place in Europe after about 1500. The new pattern of spatial interaction has continued to keep northwest Europe far ahead of the Mediterranean Basin in population and economic development.

Human society is in a state of constant change, in itself and in relation to physical distances and the environment, which do not alter. The most important reason for this second sort of change is changes in human knowledge and technical powers. People learn to do more and different things with the earth, and thus its resources and its distances affect them differently. Less tangible things, such as human wants, tastes, or cultural and ideological standards, also change. Migrations of people further help to spread change and to alter land use and other aspects of cultures or economies. The earth does not mean the same thing to people at different periods, nor does it have the same effect on them, even though some of the major aspects of the relationship may remain relatively constant in a given area. Probably the most dramatic reminder of the dynamic quality of human relation to the earth and of the impact of changing spatial relations is what has happened in North America since 1492. A land which supported perhaps one million stone-age people in 1492 now supports well over two hundred million of one of the most technically and economically advanced people in the world. California in 1492 was the home of what appear to have been some of the most primitive of all the American Indians.

One may think of any society as occupying a certain stage in the development of its adjustment to its geographic base and to its place in the spatial frame. The progression may be up or down in terms of its development relative to the rest of the world. Civilizations decay as well as grow, for instance the earliest city civilizations, so far as we know, which occupied sites in or on the margins of the great desert of Eurasia. These sites at the time were highly advantageous but now support societies marked by relative poverty and lack of economic development. It is not space or the environment which have altered, but human use of them and human ability to make more productive use of other areas whose advantages are now seen to be greater in the light of new powers and wants.

Maps and the spatial frame

The map is the distinctive tool of the geographer, for it is only with a map that spatial relations and spatial distribution can best be seen and analyzed. Direct observation of any area, even over a long period, can pick out only individual details or small pieces of landscape. Larger patterns can be perceived only when larger areas are seen as a whole, reduced to an observable scale, and many of the almost infinite details of reality eliminated in order to concentrate attention on a few matters in the large. Any map selects only a few patterns to represent, for example,

MAP TYPES

a. Land use map
b. Rainfall map
c. Cartogram
d. Computer
   terrain drawing

Figure 5 Sample spatial patterns

elevations above sea level or distribution of population. An aerial photograph shows most directly observable concrete reality, at a reduced scale in which some large patterns can be seen. But the profusion of detail and the fact that many spatial patterns are not directly observable or do not stand out in an aerial photograph make such a photograph much less clear than a selective map. A map can also represent areal traits such as language distribution, density of population, average annual temperature, incidence of illiteracy, flow of traffic, or distribution of incomes – all of which cannot be seen directly as large patterns. Most of the aspects of the earth and of human society can be mapped. Representing them on a map makes it possible to understand them better or from a different point of view, if only because they are placed within the frame of space and can thereby be related to other phenomena which have spatial form. Figure 5 shows a few examples, but see also Figure 23 on page 184. There is a growing variety of mapping techniques to represent different types of spatial form, and there are very few aspects of human society which do not have spatial form. Any atlas makes use of several different mapping techniques, and others are used elsewhere in this book.

Essentially, two kinds of jobs are performed by maps. They show spatial distribution or occurrence (spatial form), and they show relationships. The two are of course interconnected, since the nature and degree of relationships will depend in part on the amount and nature of the intervening space – how far it is from A to B and whether mountains, plains, seas, forests, farming, manufacturing, or dense or sparse populations lie between them. An atlas is designed to show as many of these things as possible in a series of maps – no one map can show them all without being illegible. Any map is thus an oversimplified and selective symbol of reality. Some distributions are best shown by a pattern of dots, some by lines, some by colors or shades, some by other symbols. Relations may be indicated by lines, such as transport routes, whose thickness varies with the volume of flow, or by the superimposition of distribution patterns to show coincidence. Routes or interconnections over space may simply be indicated by line symbols. Most relations, however, are more implicitly than explicitly indicated on maps, and they must be deduced or extracted.

A map is also a convenient inventory of selected information for a certain point in time, made available for immediate visual inspection, in the form of exact locations and spatial patterns which are manifestations of a variety of relationships. Such relationships may often be best or even first perceived through the process of making or interpreting a map and insights derived which were not guessed at earlier (see for example the discussion of China as a region in Chapter 2). The map is a powerful tool of analysis and not merely a collection of data. In addition to suggesting explanations and interrelations through its selective simplification of complex reality, it can direct attention to questions whose existence or importance were not seen before and can thus advance both theory and understanding. Anyone who uses maps should realize the varied dimensions of their usefulness and should look for more than simple locational data. Selected spatial patterns in cartographic form can raise and can suggest answers to a great variety of basic questions about the composition and interrelations of both the human and the physical landscape.

Scale [2]

Patterns of spatial form and spatial interaction can be examined at widely

Map 4 Part of southeastern Michigan, from topographic sheet. *The contour interval (difference in elevation between contour lines) is twenty feet. The heavy line across the lower part of the map is the basis for the* vertical profile *below, showing elevations along that line.*

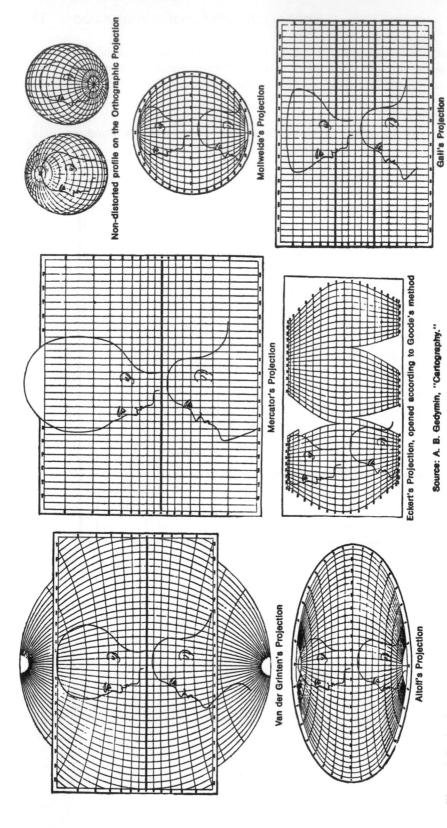

Non-distorted profile on the Orthographic Projection

Mollweide's Projection

Gall's Projection

Mercator's Projection

Eckert's Projection, opened according to Goode's method

Source: A. B. Gedymin, "Cartography."

Van der Grinten's Projection

Altoff's Projection

Figure 6 Projection comparisons

varying levels of detail. Map scale serves in effect as a filter, appropriate to the level of generalization required. It eliminates irrelevant or confusing detail and includes only what is directly useful. A map is thus not only a selective symbol of reality, but a greatly reduced image. The degree of reduction is indicated by the scale of the map, one unit on the map representing a million units on the earth, for example, or a scale of 1 to 1,000,000. A map of the United States alone at this scale would nearly cover the floor of a small room, however, and maps of large countries or continents are therefore usually drawn at smaller scales like 1 to 20,000,000 or more, or of the world at 1 to 50,000,000 or less. ("Small" refers to the fractions; in these terms, 1 to 50,000,000 is a smaller scale than 1 to 1,000,000, although it may be used to show a much larger area.) Naturally, the smaller the scale, the less detail can be shown. A map with a large scale – like 1 to 62,500 (often used in the United States because it is an easily divisible fraction which is nearly the same as one inch to one mile or 1 to 63,360) – can show individual buildings, as in Map 4. For detailed analysis of small areas, such large-scale maps are essential, but most atlas maps are necessarily drawn at smaller scales. It will be helpful to examine the variety and progression of scales in an atlas and the bar scale commonly used; this is a line representing miles, kilometers, or other units on the scale of the map, which is a great convenience in measuring distances, as in the sample in Figure 3. Ideally, all maps should carry an indication of *scale*, a *date* (preferably referring to the date of the information they portray as well as date of publication), a *title*, an *orientation* based on compass points (usually in the form of an arrow indicating north), a *legend* or *key* explaining the symbols used, a *grid* showing latitude and longitude (see below), and for larger areas an indication of projection (see below).

Distortion

Despite their great advantages, maps must be used with caution because they are necessarily distorted representations of the earth as well as greatly reduced and oversimplified selective symbols. The earth is spherical and three-dimensional; a map is flat and two-dimensional. Long habit of looking at flat maps tends to make one forget that it is impossible to represent a spherical object or any part of its surface on a flat plane without distortion. The simplest proof of this is to lay flat a whole orange skin, or the two halves of a rubber ball. It cannot be done without tearing or stretching. A map distorts reality in shape, distance, direction, or total area. Most maps do one or another of these things, depending on the method used to project the earth onto flat paper. A few samples of *projections* and the distortions they produce are shown in Figure 6. Most atlases include some projections which distort shape and some which distort area, each useful for different purposes. Distortion is, of course, maximized in maps of the world as a whole (which cannot really be seen as a whole at all, since one cannot see around it), barely noticeable in maps of areas the size of the United States, and negligible for smaller areas. The distortion of small-scale maps also means that distances or directions are often shown inaccurately, and care must be used in measuring distances, especially on projections which do not show the true size of areas. The only approximately true representation of both shape and area is a *globe*, and flat maps should be used only with that model in mind.

Latitude and longitude

For purposes of measuring and indicating exact locations, a grid of lines is superimposed on the map, as in Figure 7. The parallel lines extending east and west are called parallels of latitude, both north and south of the equator (which is midway between the north and south poles). These parallels of latitude are intersected by lines extending north and south,

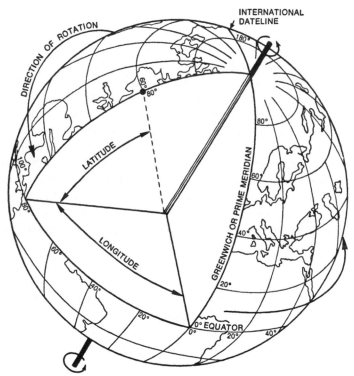

Figure 7 Latitude and longitude. *Cutaway globe illustrating the angular relationships of meridians (longitude lines) and parallels (latitude lines). The point lies at 60° north latitude and 80° west longitude.*

called meridians of longitude. Meridians are not parallel, because each of them connects both poles; they therefore converge toward the poles and are most widely separated at the equator. One of these meridians of longitude is arbitrarily chosen as the base point of reference, or the prime meridian. For convenience and by established practice, the meridian which runs through the Royal Observatory at Greenwich near London is generally recognized as the prime meridian, although any other one could just as well be used provided everyone agreed on it. The fact that Greenwich has been generally accepted reflects the prominence of Britain in the modern world and the pioneer work of British scientists, explorers and map makers. Longitude is thus measured in terms of distance east or west of the prime meridian of Greenwich, and latitude in distance north or south of the equator. The earth is divided for these purposes into 90° of latitude and 180° of longitude, 90° north and south of the equator, and 180° east and west of Greenwich. The degrees are further subdivided into minutes and seconds for greater accuracy. Thus any location may be stated as so many degrees, minutes, and seconds north or south latitude and so many degrees, minutes, and seconds east or west longitude. By such means it is possible to fix locations very accurately and to pinpoint a house or a ship, for example, within a few yards.

180° east of Greenwich and 180° west of Greenwich are of course the same thing. Where they meet another line must be drawn, in effect to separate east and west as the equator separates north and south, and also to mark the change on the calendar from one day to another, in accordance with the rotation of the earth. Passing eastward from one time zone to another, clocks are advanced one hour, but passing eastward across the line of 180° of longitude, one enters the preceding day. Travel eastward, toward the rising sun, has progressed so far by 180° that it has caught up with and overtaken the day which arbitrarily began at 180° twenty-four hours earlier. Crossing 180° in a westerly direction, the reverse is true; clocks are set back from one time zone to another westward, but at 180° one enters the following day, which is arbitrarily beginning at 180°. The change of date at 180°, the International Date Line, often confuses people; it is best to regard the day as beginning and ending at 180° so that the crossing of this line changes the date on the calendar.

Maps have been discussed in this chapter because they are an essential means to the study of space. Such study is only incidentally concerned with where individual things are and more importantly with how physical features, man, and his works are distributed over the earth, why these spatial patterns have developed, how they are interrelated, and how and with what affect spatial interaction takes place in all of man's activities. The map is both tool and symbol of geography's concern with spatial form and spatial relations.

NOTES

1. Peter Heylyn, *Microcosmus, or a Little Description of the World* (London, 1621), p. 11. (Note the use of the literal translation of "geography.")
2. Techniques for showing relief and some of the problems of scale are discussed in Chapter 12 on land forms.

4 SPATIAL SYSTEMS

Geographic analysis assumes that phenomena are distributed over the earth, or within any smaller part of it, according to one or more systems of arrangement which are at least potentially identifiable. This important aspect of geography is concerned with the spatial ordering of things and especially of human activity. Many of these spatial systems are more or less obvious: political units such as national states or counties, patterned systems of land use, such as rubber plantations or market garden areas around cities, the placing of houses along streets, and the arrangement of rooms and furniture within each house. Some are less obvious, such as the distribution of income, the incidence of particular diseases, or the spread of ideas. But whether obvious and directly observable or not, spatial patterns can be fully understood only when they can be delineated as individual or collective systems resulting from a combination of factors which can affect spatial behavior. Some patterns of spatial distribution are random or haphazard, but this situation is quite rare; nearly all distributions are non-random, and are shaped by particular factors, although the relevant factors may not always be easy to distinguish. The random/non-random distribution is, however, important to recognize in any study of spatial systems.

Spatial behavior

As suggested by the examples given above, spatial arrangement is evident in the entire range of human activities, beginning with a phenomenon which has come to be recognized as "personal space." Each individual orders his or her activities within or according to certain spatial limits and preferences. This ordering includes such things as the placing of furniture, the layout of a house, and even the ordering of the contents of a bureau drawer, a desk, or a kitchen cupboard. Some individuals are more sensitized than others to spatial arrangement; some are said to have "a good sense of direction," and it is often these same people who order their personal space coherently. Others have a poor gross sense of direction and often have difficulty finding things even within the smaller and more personal scope of a house or a room where, from their point of view, spatial arrangement is not coherently ordered. Some people dislike crowds or small confined spaces; others enjoy and even seek such situations.

There are, however, patterns of personal and interpersonal space which tend to be consistent within any given culture, and to contrast with similar patterns in other cultures. One of the commonest of such patterns relates to interpersonal contact. In any given culture most people tend to accept or to require a certain minimum distance between themselves and others with whom they may converse; they also recognize and use a common set of gestures and other non-verbal signals often referred to as "body language." Each culture may have its own requirement of how large this interpersonal distance should be and may have its own, quite different and accepted, vocabulary of "body language." This concept is related to what is sometimes rather misleadingly referred to as "territoriality," a concept derived from the study of behavior of other animals where the notion is perfectly appropriate even though it may not fit human behavior as well. Most species of animals and birds are dependent for their survival on the control or exploitation of defined pieces of territory which can provide them with food or ensure their ability to reproduce; they therefore attempt to defend this territory against trespass or use by others of the same or other species and may also stake claims to it by various means or signals. Bird songs, for example, are at least in part a means of establishing claims to a given foraging and nesting territory. Human survival problems are different, primarily because we have passed the hunting-and-gathering stage and are able to assemble what we need over great distances. Nevertheless, there is an element of territoriality still involved in individual perceptions of personal space, and, by successive extension, in perceptions of family, home, or group space as well as of national space. Perhaps the major difference between humans and other species in this respect is that animals have long ago learned how to order potential conflicts over spatial arrangements and allocations, whereas people still engage in both individual and mass destruction in pursuit of a territoriality which is not adequately provided for or regulated.

Spatial perception

Each culture has its biases, and its own idiosyncratic perceptions of space. Most individuals, and most culture groups, have a spatial perception which is both warped and incomplete. You have probably seen examples

Figure 8 A New Yorker's view of the United States

of cartoon maps, or cartograms, made to illustrate this point, for example, the New Yorker's view of the United States (Figure 8) or the American tourist's view of Europe. In these the viewer recognizes the existence, location, importance, or size of something in accordance with his or her perception of what is important. Individuals have a similarly warped and incomplete *mental map* even of the much smaller space which they occupy or move around in. Reality is immense, and what it is possible to notice, remember, and organize is only a small selection, based on what is considered important. On a daily journey to work along the same route, a commuter will miss (that is, sort out of conscious attention) most of what is directly perceivable, and will mentally order that particular space corridor in accordance with personal criteria, which, for example, may well reflect travel time rather than distance. He or she may be totally ignorant of what lies immediately outside the space corridor through which the daily journey to work moves but may be familiar in detail with other more remote areas or have an accurate impression of their location and nature even without having gone there. Culture affects individual perceptions of space in consistent ways because it tends to define what is important, and is also the principal medium through which individuals learn or acquire information. North American culture is relatively self-contained, isolated, and self-satisfied, but the people are closely aware of western Europe, although their information about Europe is scanty and often inaccurate. Much of the rest of the world is perceived even more vaguely, if at all, as many sample studies and questionnaires have demonstrated.

Such cultural biases are perfectly natural and universal. However, as travel time shrinks and all parts of the world become increasingly interdependent, or as conflict threatens as a result of imperfect knowledge or understanding, warped and incomplete perceptions of global reality become dangerous. There is less excuse for geographic ignorance on the part of North Americans, or for the dangerous biases which their culture tends to produce, than for similar attitudes on the part of less wealthy populations in the past.

Settlement patterns

Perceptions of space vary over time from place to place or culture to culture, and among individuals. But concrete phenomena and actual flows are arranged in space non-randomly and according to identifiable systems; it is part of the geographer's job to identify and analyze them. Settlement patterns make a good illustration, and are perhaps the most widely studied forms of spatial systems. Cities, towns, and villages grow up at certain locations and perform certain functions primarily as a result of spatial interaction, as discussed in the preceding chapter. The nature and amount of the settlement's interaction with other places will largely determine its size and its own nature and will even influence its internal spatial arrangement. Modern Western cities have exploded in size during the past century in particular, as the means of transport have multiplied and it has become possible to move larger amounts of goods more cheaply over greater distances and at greater speeds. This enables the city to perform more functions on behalf of larger areas, to bring in the food and other supplies it needs to maintain its own population, and also to draw its own work force from greater distances so that the urbanized area (including commuter suburbs) has rapidly extended as cars, buses, and trains bring larger areas within bearable travel time from the city. (See Map 5.)

Origins of commuters to Chicago CBD   1 9 8 0

North Shore Radial

Northwest Radial

Oak Park–West Radial

Burlington–West Radial

Illinois Central–South Radial

Map 5 Origins of commuters to Chicago. *Each dot represents 1000 commuters from that area to the central business district (CBD).*

The arrangement of agricultural villages reflects the same kind of spatial constraints or circumstances, even in areas where mechanized transport is absent. Where movement is by foot, animal power, or water, the spacing of large settlements will reflect most importantly corridors along which bulk goods can be moved cheaply, primarily waterways or sea coasts. Smaller settlements tend to be spaced in accordance with local agricultural productivity and to be closest to each other where productivity is highest. This pattern is seen in most of Asia, for example, where the distribution of villages on an agricultural plain looks like the sketch in Figure 9a, based on an aerial photograph of a small part of south India.

In these situations, productivity per acre is high and fields are small. They are grouped directly around the village, and the farm workers from each household are easily able to walk to their plots every day. The size of the village is effectively limited by the number of households which the

Figure 9a
Villages on a south Indian Plain

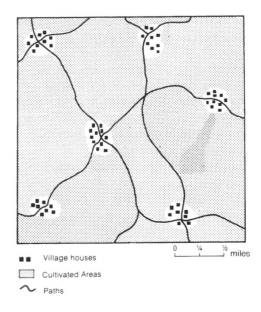

■■ Village houses

▨ Cultivated Areas

∿ Paths

0   ¼   ½ miles

Figure 9b
Standard Asian village pattern

■■ Village houses

◁ Cultivated Fields

▦ Uncultivated Uplands

∿ Paths

0   ¼   ½ miles

area can support on the basis of agricultural land within easy walking
distance, a measure which is of course determined also by the productivity of the land. As shown in Figure 9a, an agriculturally productive
landscape tends in this way to be broken up into groups of small fields

clustered around villages that are relatively close together, with the outer borders of the fields of one village adjacent to those of the next village. In a less productive setting, village spacing would be wider. Like most spatial systems, these patterns reflect also the *principle of least effort*, whereby things are arranged so as to minimize movement, labor, or expense. The same principle can be seen in the spatial layout of modern cities (see Maps 5, 6 and 22, for example, and Figure 10), and in the arrangement of agricultural land uses at varying distances from a city, as illustrated in Figure 11.

Spatial arrangements within each village reflect their function, and their spatial constraints, as shown in the sketch in Figure 9b. Houses are closely clustered along a central or single path, and fields radiate out from the village on all sides. In such a situation, people necessarily live closely together, and various forms of cooperative effort are involved. Relations with other places and other people tend to be minimal, but within the village relations and interdependence are close. These circumstances are in turn reflected in the traditional social structure characteristic of Asia: an exceptionally closely knit family system, a firm web of intra-village connections, little scope for individual as opposed to group interests or values, and a generally closed or self-protective face to the extra-village world. All of these characteristics have clearly been influenced importantly by the spatial system, the particular frame in which village Asia existed and to a large extent still does exist.

Urban patterns

The contrasting spatial circumstances of the modern Western city are clearly related to its contrasting social and cultural characteristics, and to many of its contemporary problems: individualism, alienation, rootlessness, mobility, and rapid change. Modern Western cities are also primarily commercial phenomena, and their internal spatial patterns and forms are shaped to a major extent by their commercial/industrial functions and by the commercial value of urban land. The layout and architecture contrast with those cities whose chief functions are ceremonial, symbolic, or administrative, including most cities built before the nineteenth century. In most contemporary American cities, land values tend to reach a peak in or close to the city's central business district but to fall off away from the center in an irregular pattern. Large space users (railways, most factories) must obviously seek cheaper land, but functions whose success depends on easy access in high volume, especially department stores and other retailing, must seek the center or outlying centers where transport access is maximized. They tend to cluster near points where pedestrian traffic is greatest or where individual customers can most easily reach the area by car or by public transportation.

With the advent of the private car and the consequent decay of public transportation in many cities, department stores and other retailing, formerly concentrated largely at the center where public transport lines

MODELS OF URBAN SPATIAL STRUCTURE

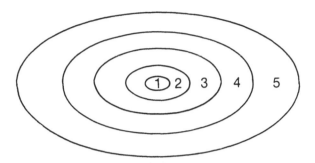

1. Central Business District
2. Deteriorating Transition Zone
3. Workers' Homes
4. Middle Class Suburbs
5. Commuters' Zone

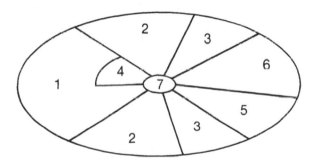

1. High-rent Residential
2. Intermediate-rent Residential
3. Low-rent Residential
4. Education and Recreation
5. Transportation
6. Industrial
7. Core

Figure 10 Models of urban spatial structure

converged but where the private car was at a disadvantage, have had to move or to establish stores in outlying locations where car access is better and where ample lower cost space is available for huge parking lots. For most cities, the effects of distance from the central core are however evident in decreasing intensiveness of land use, as shown in the following diagram:

Modified after Von Thünen

ZONES OF LAND USE
1. Non-agricultural urban area
2. Poultry, horticulture and market gardening
3. Market gardens, fruit and fluid milk
4. Dairying, truck farming and other types of intensive agriculture
5. General farming, grain, hay and livestock
6. Open range or forest-range in areas of semi-arid plains, forests in moist hilly areas
7. Minor grazing and dairying, forestry on steep slopes

(Lower half of figure suggests modifications induced by variations in landscape.)

Figure 11 Hypothetical land use zones around a city as a function of distance

On a wider basis, distance from a major urban center usually plays a role in influencing what rural areas will be used for. Figure 11 attempts to show in generalized form the effect of market distance on agricultural land use, eliminating all factors but distance and using lines instead of the overlapping transition zones which are closer to reality. Even in subsistence farming systems, land nearest the village will be worked most intensively, and the village will tend to be located in the optimal soil and water areas. Land farther out will not repay intensive cultivation so well. The von Thünen model shown in Figure 11 may thus apply in such situations also.

Within any city, land values are also reflected in the pattern of residence, including its segregation by race, cultural origin, and occupation, since these factors are related to income. Black residential patterns make the clearest case because, at least in most American cities, Blacks are the largest easily identified separate racial and cultural group which also is largely restricted to low-income occupations. Boston's Chinatown is shown in Map 6, where the zones which surround the Chinese area have made it less desirable for residence and kept housing and land costs relatively low, a situation taken advantage of by disadvantaged groups.

Map 6 Land use in Boston

Spatial association and migration

However, land values and occupation/income are not the only factors which affect residential clustering by racial or cultural groups in most cities. Much depends on the less easily measured influence of cultural ties. Culture varies, of course, even within a single national group or single city; it is affected, for example, by occupation, by income, by sex, by age, and by other similar factors which also help to form interest groups sharing certain attitudes, values, or circumstances. Members of a suburban bridge club or tennis club, for example, may have relatively little in common and minimal interaction with a central city factory workers' bowling club and will have different points of view on most subjects. Even their speech will differ to some degree, as well as their preferred diet, dress, and musical tastes. Culture may indeed be as important, or more important, than race in influencing people's choices of interpersonal association and interaction, which in turn help to determine residence patterns. Clearly discernible racial differences, such as skin color, do unfortunately greatly sharpen distinctions in the minds of many people and lead to (often unfounded) stereotypes about the supposed behavior of supposed racial groups. Perception of differences may lead to segregation imposed by the dominant racial or cultural group, as is regrettably the pattern in most American cities, but it may also tend to strengthen the sense of group solidarity on the part of those who are discriminated against. They may have in common not only a set of cultural traits and thus be inclined to seek one another's company but may also share common oppression. Both sets of factors tend to affect residence and interaction patterns.

Such voluntary association may especially appeal to groups or individuals who are relatively recent arrivals in a city or a national unit. As strangers who also suffer to some degree from linguistic, other cultural, or economic handicaps in competing or interacting with the existing dominant group and who may in addition be discriminated against, they understandably tend to stick together. This pattern has been seen, for example, with Spanish-speaking Chicanos and with Asians in the United States and with Black migrants to northern cities from the south, all of whom have tended to settle in quite sharply segregated patterns associated with low-value (but often high-rent) areas close to the center of the city, where some public transport is available but where manufacturing, wholesaling, or the more general urban blight which has grown around the central business core of most American cities make these areas undesirable to more advantaged groups.

The study of migration patterns: what kinds of moves have taken place, which groups and individuals have been involved and over what routes, and what kinds of factors affect this process, is a rapidly developing field of analysis in which many geographers are involved, since it is an obvious aspect of the spatial dimension of human society. Migration affects, of course, not only the migrants themselves but the areas to which they move, and those which they have left, whose character also changes, in some cases fundamentally. Most migration takes place under some form of duress or pressure, usually economic need but unfortunately often also discrimination or, more severe, oppression. All may however be affected by individual perception, economic need particularly so since it is of course a relative matter. Before the individual can rationally decide to move, he or she must conclude that circumstances elsewhere are in some way better. His or her information about other places may be, and usually is, imperfect or even erroneous, and for most people migration is a painful and difficult enterprise not to be undertaken lightly.

The distribution of population is of course itself a spatial pattern, and is discussed as such in Chapters 7, 9, 10, 11, 12, and 13 as it is related to the different elements of the natural environment which together form the basis on which man lives. Population is discussed in greater detail in Chapter 13, in its relationship to resources and resource use, and in terms of the problems attendant on the explosive increase in human numbers in the past century, an increase still continuing. However, population pressure has long been a direct factor in migration patterns. In general, most of the great migrations of the past, and through perhaps the first quarter of the twentieth century, have been in response to economic pressures, especially famine, or to wars. Such movements are still from time to time taking place – for example in Southeast Asia – but, especially in the so-called developed or affluent world, most migration has become increasingly a matter of individual choice and often takes place within a given national unit, in search of still greater economic opportunity, or of such amenities as milder climates. Mexican immigration to the United States is however a good example of movement resulting from population pressure and unequal economic opportunity across international borders.

Diffusion

Most potential migrants base their decision on information about other places derived from word-of-mouth reports from other migrants; these reports are usually diffused (spread) through a chain of intermediaries and hence may become distorted or may be unrepresentative in the first place. The whole process of information diffusion, an important dimension of every aspect of change or innovation in human behavior, takes place over space and hence is properly a subject for geographical analysis, as is the diffusion of any change. Practically all aspects of technology and culture and most plants, animals, and diseases were once confined to relatively small areas or were initiated in a particular location and have since spread more widely or globally.

The timing and the means by which these patterns of diffusion take place have become one of the fastest growing fields of investigation in the past thirty years within the discipline of geography, concerned as it is with the spatial dimension. Human migration is merely one form of spatial diffusion, but one which has been happening since the emergence of *Homo sapiens*. The dissemination of crop plants, such as wheat or potatoes, and the unplanned spread of wild plants as unintended human baggage may be almost as old and is still continuing. In most areas of the world, relatively few of the cultivated plants or of the wild species ("weeds") which invade fields, cleared areas, and roadsides are originally native to that area and many have been introduced very recently; the same applies to a surprising number of wild as well as domesticated animal species. Perhaps the clearest illustration is provided by pheasants and chicory (a common roadside or field "weed" with lovely light-blue flowers) in North America. Both of these are now almost universally diffused in this continent, but both were introduced here originally in very small numbers in the second half of the nineteenth century. Two of our commonest birds, the English sparrow and the starling, were also introduced, in a few pairs each, in the 1870s. Horses, cows, pigs, sheep, and chickens were all introduced into the Western Hemisphere in 1492, as they had been into Europe (from Asia) some millennia earlier. With them, and with the cultivated plants introduced, came animal and plant diseases, as human diseases spread with human migration.

With the greatly increased human interaction and actual movement of peoples and goods attendant on improvements in transport in the past few centuries, human diseases have been able to spread much more widely and rapidly. Diseases seem also to follow a cyclical pattern of incidence and virulence, pulsing in spasmodic waves over time and, with the newly increased means of diffusion, over space. The frightening increase of human population has also created better conditions for the spread of disease, through crowding, through eroding the nutritional or other economic basis of subsistence, and simply by providing more hosts. In time, most populations acquire a degree of immunity or at least resistance to many diseases which are endemic in the area they occupy. New diseases, imported from another area, may be totally devastating, as has happened repeatedly in the course of human migration. The common cold, for example, was a dangerous and occasionally fatal illness to some North American Indian and Eskimo groups when it entered the continent with early European settlers; measles, which they also brought with them, brought death to the great majority of the local population who contracted it. Only widespread intermarriage (or interbreeding), which in effect transferred European genes to the local population and hence diffused a degree of resistance to these and other imported diseases, prevented an even greater decimation.

Most attention has been paid in the rapidly developing study of diffusion to the spread of technological and cultural innovation, which is indeed a fascinating process, in part because one can see it operating and because it is continuing to take place in new forms on a short time scale. Most innovations are not universally diffused and most are spread via information networks which form distinct spatial patterns and follow specific routes. Especially in the early stages of spread, the innovation may appear only along certain routes or within certain kinds of areas, skipping over intervening areas.

Systematic study of the contemporary or recent diffusion of innovation has so far dealt with a variety of examples, including ownership of radio or television sets, telephones, and automobiles, the spread of new crops and of new cultivation techniques or machines, and the changing spatial pattern of attitudes (political, ecological, racial, religious), clothing styles, dietary innovations (for example, pizza), drive-in banks, and saunas, to name only a few examples. Each of these innovations tends to depend for its spread on a somewhat different chain of circumstances and a different network of diffusion, including of course the means by which information about the innovation is made available to potential adopters and the degree to which potential adopters are influenced by information or example from particular sources. Each innovation is likely to have its own discrete pool of potential adopters, many of them not necessarily susceptible to other forms of innovation, or in any case less likely to be in touch with a different interpersonal or intergroup network which might stimulate them in favour of another innovation. Physical proximity does of course assist the spread of innovation from one adopter to another, but commonality of interest or circumstances may be equally or more important. Wheat farmers, for example, are more likely to be directly or indirectly in touch with and to adopt innovations from one another or from a common network such as an extension service, even though they may live considerable distances apart, than they are from their immediate pig-raising or school-teaching neighbors. Other more general innovations, such as telephones, are likely to spread through a series of discrete networks and on a different time schedule: first perhaps within a given income or occupational group in a city, then within similar groups in other cities, then to other urban groups, and finally to a series of rural groups, each constituting a separate network; or the innovation might diffuse primarily through income or occupational networks regardless of urban/rural distinctions.

Spatial inequities

Whatever the changes and tendencies toward uniformity brought about by the spatial diffusion of innovation, however, the world remains enormously varied, both in its regional range of physical diversity and in the way in which people live. This is not merely a matter of regional culture, preferences, and attitudes but involves gross disparities and inequities in the material bases of life. Two-thirds or more of the world's people, sometimes euphemistically called the "developing" world, live and die close to and periodically below the margin of subsistence, while the remaining one-third enjoy increasing affluence. People have not yet learned how to diffuse the economic abundance which their new technology makes possible. Most of that new technology is used instead to strengthen the power and wealth of those who already have it and to support international rivalries and conflicts aimed at the same objective. Relatively speaking, the poor grow poorer and the rich richer as the gap between the "developed" and "developing" sectors continues to widen. These inequities are overlaid into a pre-existing pattern of racial and cultural differences. A series of circumstances of very recent origin has during the past two centuries – a brief moment after all in the course of human history – brought wealth and power mainly to the people of Europe and North America (plus Japan), most of whom share a common racial and cultural grouping; the rich of the world are predominantly white. It is not easy for the rest of the world to avoid resentment, and to use pre-existing racial, religious, linguistic, and other cultural differences as symbols of rivalry or as divisive rallying points.

These kinds of spatial patterns and the factors which account for them are clearly the most important of all. The geographer's efforts to identify and analyze the spatial dimensions of human society need to be concentrated on those which help to divide mankind in these destructive ways, and on the search for answers which can help to ensure that differences, which will always exist, do not lead to injustice or to conflict. We are beginning to learn a good deal about spatial patterns of difference and interaction. That developing knowledge needs to be used.

5 THE ENVIRONMENTAL FACTOR

The physical environment means to a geographer everything except people and their works: climate, soils, vegetation, wild animal life, land forms (mountains and plains), the atmosphere, the oceans, minerals, and water. Geographers call these things the *environment*, or literally the *surround*, because they are concerned with the human factor. The word "environment" has meaning only in reference to some organic unit existing within it for which it provides the matrix or setting. It is obvious that the physical environment may have a profound effect on people and on how they organize their activities. But it is appropriate to speak of an *inter*relationship because people in turn act on their environment and alter both it and its effects on them. The environmental influence is thus far from being a constant, as human society itself is not constant.

Human modification of the environment

People change their environment in a great variety of ways and have fundamentally remade the landscape of every area of the world which they have occupied in large numbers. In many cases succeeding periods of human occupance of the same area have each brought about a different change in the landscape, drained or irrigated it, deforested or planted it, enclosed it or left it as one great common. People usually remove the forest first, although even this has not yet been accomplished in the taiga forests of northern Asia and North America or in most of the equatorial rain forest. Deforestation probably has more widespread effects on the rest of the environment and in turn on people than anything else they do. Agriculture usually follows deforestation and basically changes the soil or may seriously deplete it, and also revolutionizes the appearance of the landscape. Depending on the nature of the agricultural system, woods and clearings are replaced by hedgerows and plowed fields, fences of varying kinds in each region, terraced rice fields, orchards, plantations, irrigated basins, vineyards, truck farms, pastures, open ranges, artificial lakes, dams, wells, ponds, or fields of grain. People also build a more or less dense network of roads and later railroads and airfields over the landscape, dig canals and dredge harbors, dig mines and oil wells, use or exhaust mineral deposits, and reap a harvest from the sea, changing it as well. All these changes in the environment react in turn upon people and influence their settlements and activities. Because we have altered the environment and learned to use it differently, we have thereby come under the influence of a variety of new factors – some positive, such as new resources, but some negative or limiting, such as soil erosion – which were unimportant before or were important in different ways. Finally, villages and towns grow into cities which in the modern period may cover many square miles with paved streets, factories, shops, and dwellings.

Where "nature" has been left, it has mostly been modified, at least in the environmentally favorable parts of the world where people have multiplied. Even the undestroyed mid-latitude forests are not left alone but are systematically managed or replanted. Few parts of the modern earth outside the polar areas or the equatorial tropics would be recognizable to Neolithic or even to classical or medieval people. Animals kept or encouraged by humans also play a part in modifying the environment. This is especially true of the grazing animals, and most of all of sheep and goats, whose close cropping prevents the regrowth of trees or bushes and even sometimes of grass. Overgrazing leads to serious erosion and may thus produce a change in the effective climate; many desert-like areas are in fact the work of goats. People have been able to upset seriously the complex balance of interdependence among the biotic elements (animals, plants, birds, fish, insects) and the rest of the environment.

The force of environmental factors in their influence on people has led some investigators to give them undue weight, as suggested in Chapter 1. As usual with such deterministic approaches, there is a core of validity in the environmental approach and an idea of great usefulness, provided it is used intelligently. The physical environment is one of many factors which influence people. Such political considerations as subsidies, tariffs, central planning of the economy, or differences in the size of markets as between different political units may also be important or even predominant influences on agricultural systems, manufacturing patterns, or trade flows. Such cultural factors as taste preferences, traditional systems and techniques of agriculture or grazing, or systems of social organization may be more important than the physical environment in determining the type of land use or the nature of the economy. Just as important, human technical abilities and wants are continually changing, and for all of these and other non-environmental reasons, the human use of the earth varies enormously from place to place and century to century.

Similar environments are used differently, different environments are used similarly, at different times and in different places. Similar cultures may make similar use of different environments. Western Europe, south-eastern Australia, and eastern North America tend to look alike, despite the differences in their physical environments, because they have been settled by people with similar cultures and similar techniques. By contrast, southeastern China and southeastern United States look very different, despite the close similarity of their physical environments, because of the great cultural and technical differences between their respective inhabitants, reflected in their use of the land. Chinese culture traits and associated patterns of land use have also spread into other parts of Asia and have created there what might be called a Chinese landscape. The Europeans, when they became active agents in Asia, made a different use of the land and created a distinctive landscape in the form of plantations, as well as building cities which are European in appearance. Even in so rigorous an environment as the desert there are pronounced differences in culture and economy between arid areas which are physically very similar. For example, the deserts of central Asia, Australia, South Africa, and southwest United States all support different people with wide divergences in their culture and method of land use. Different people look for different things in the same environment and produce quite different results.

Determinant or influence?

These obvious facts and the continuing changes in human use of the environment in any one area invalidate the theory of environmental determinism. It is a tempting theory because it purports to explain the complexities and variations in human society in terms of a non-human constant. The physical environment may well be relatively constant when compared with human society, but its effects on people are far from constant and often far from clear or direct. Historians have remarked that the only thing which does not change is change itself. However, even this statement is not entirely true, since the process, nature, speed, and direction of change frequently vary widely, and to a large extent reflect of the over-all growth or change of human society. Such an analysis applies also to changes in the environmental influence on people.

Nevertheless, the influence of the physical environment is profound. Few aspects of human society are free from it, although its effects may be disguised and are not often predominant. There is a risk that in discarding the theory of environmental determinism the baby may be thrown out with the bath water, as happened to some extent when the unsoundness of environmental determinism was more clearly realized. Many scholars abandoned all attempts to consider the environmental influence or even to consider causality in general and confined themselves to simple description of spatial patterns and regional units without attempting to interpret or relate what they described. This was a serious loss; the purpose of study is to understand and explain, or at least to ask *why?*, not merely to accumulate data. Avoidance of the environmental factor also neglected a major function of geography, which is to set people within the frame of the earth which they inhabit; this necessarily involves critical examination of human interrelations with their physical base, as well as the variety of other factors which influence them and which contribute to regional differences. Fortunately, the discipline of geography has increasingly realized the central importance of that task.

"*Determinism*" as an approach has been replaced by "possibilism" which assumes that people have a large degree of choice even though the environment may set certain limits on what they can profitably do. Of the many things which are possible or which environmental conditions allow, people choose to do certain things in accordance not only with their own inclinations but as a result of a variety of other influences or factors which help to shape their civilization, such as political pressures, traditional forms, cultural values, or the development of particular techniques. Environmental conditions may tend to make one or a few choices most attractive or most successful, but choices may differ widely among different societies and at different times. The rejection of determinism is not merely a belated recognition of human free will, but more importantly a realization that human society and its choices are extremely complex and that they are continually changing. No one factor or influence can hope to explain this complex reality; what is observed is necessarily the result of a concert of interacting factors in which the human side is fully as important as the physical environment. But this is not to say that the environmental influence can safely be ignored. Its effects can be traced, sometimes with striking prominence, sometimes as a subdued note in the concert of influences, in most aspects of human society and in its differences over time and over space.

The nature of the environmental elements and their relation to people will be left for more detailed discussion in Chapters 7 through 13, where each can be considered separately. For the present it is enough to say that there is a fundamental relationship between people and nature and that geography attempts to delineate it, in its proper place, as part of the study of spatial patterns and the analysis of areal differences in human society.

Cities and the environment

Human settlements are not distributed over the earth in haphazard fashion but in a recognizable pattern, or in association with factors which may make a particular site or sites advantageous for settlement. Since cities exist in order to provide services for an area dependent on them, they flourish at locations which maximize access. If the landscape were a productive, flat, undifferentiated plain without rivers or local differences of any kind, distance would be the only factor and the city would be located in the exact center of the area which it served. But landscapes are never entirely uniform. There are local differences in topography, soils, ground water, economic productivity, culture, political pressures, and many other factors which tend to make certain sites advantageous and which may in particular make access easier. Cities frequently grow up near points where two different physical areas come together, such as mountains and plains, land and water, forest and prairie, or desert and humid areas. Study of an atlas will show how frequent this association is, and an example of it has already been examined in the case of Timbuktu. Cities are centers of exchange, and exchange implies difference. Where differences meet is a convenient place for exchange; transportation is minimized in both directions. In moving from one kind of physical area to another, it may also be necessary to change the mode of transport, as at Timbuktu.

Cities of the fall line

From Providence, Rhode Island, through Montgomery, Alabama, lies a string of cities arranged in a line roughly parallel to the Atlantic coast but in most cases removed from it and at the inland edge of the coastal plain where the plain merges with the foothills of the Appalachian Mountains (see Map 7). Along this line the physical conditions of the coastal plain

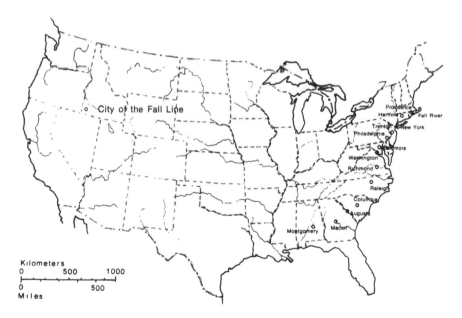

Map 7 Cities of the fall line

change and are replaced by those of the uplands or the Piedmont, as it is called (literally, "foot of the mountains"). Among the cities which punctuate the line are some of the largest in the world, and most of the others are of more than local importance. The list includes Providence, Fall River, Hartford, New York, Trenton, Philadelphia, Baltimore, Washington, Richmond, Raleigh, Columbia, Augusta, Macon, and

Montgomery. Why are they there? What are the advantages of this zone for access and exchange, the basis of nearly all cities?

A median position between two different physical regions involves many advantages. But there is also a more specific localizing factor which is related directly to the local environment. The piedmont is composed of relatively hard rock, the coastal plain is relatively soft or sandy, and the rivers flowing from the mountains to the sea have cut into the coastal plain along this line more deeply than into the hard piedmont rocks. Nearly every river along the line is broken by falls or rapids as it tumbles down onto the plain, or at least navigability sharply decreases, and above this point navigability is much less than below it. Below the falls many of these rivers, especially north of Richmond where the largest cities are, widen out markedly because they have in effect been "drowned" by an invasion of the sea and are really estuaries or bays rather than rivers, deep and wide enough for ocean-going ships.

The "fall line," as it is called, offers four specific advantages as an urban site: (1) local water power; (2) the lowest point at which the river can easily be bridged or crossed and where land transport routes are thus bunched together; (3) the point at which goods arriving from the sea must be unloaded into a different carrier, either by land transport or a smaller river boat; and (4) a place of vantage on a river which gives access to a tributary hinterland lying in and behind the mountains through which the river's valley cuts. [1]

The first two advantages are particularly important in the early establishment and growth of a city even though they may lose their importance later on. London, for instance, grew up as the lowest practicable bridge site on the Thames, and it remains so. Fall River, as the name implies, based its early industrial leadership on local water power, as did Trenton. But it is the last two of the four advantages which give a sustaining basis to a city, because as economic development proceeds, they increasingly emphasize the city as a focus of exchange. Wherever carriers must change, the process of exchange is most easily concentrated. Goods must be unloaded from one carrier and loaded onto another. They may be stored before moving on, or they may be processed or manufactured. In any case, such a point is a convenient collection and distribution center. These break-in-bulk points, as already discussed, are admirably illustrated by the cities of the fall line. The break comes there rather than on the coast because water transport is cheap and is therefore used as far inland as possible, because location in the zone between plains and hills makes the fall line city a more convenient exchange center, and because the river itself, or its valley, provides an easy route into the interior.

Usually one city in a region has the best combination of advantages and becomes much the largest. New York is a convincing example; although the fall line or the head of ocean navigation is farther up the Hudson, New York's magnificent natural harbor on the sea itself is a major asset. But the advantages of the fall line are enough to support many other cities, too, including Philadelphia, only about ninety miles from New York but still among the twenty or so largest cities in the world. Like New York, Philadelphia is not supported directly by the fall line. On the Delaware River the sharpest physical break comes further upstream at Trenton, in the form of rapids, although Philadelphia was the head of ocean navigation until the recent dredging of the Delaware above the city enabled large ore boats to reach the new steel plant at Morrisville, just below Trenton. The Schuylkill, a smaller river flowing into the Delaware at Philadelphia, has its falls within the city, however, not far from the central business district, and the Schuylkill valley provides an easy route to the west, including important coal-producing areas. Philadelphia is supported by the market and productivity of a large inland area which the advantages of the fall line help to make tributary to the city. Actual distance is, of course, an important part of access. New York had the advantage of being more or less in the middle of the most populous and productive part of the coastal plain. Philadelphia was close to the middle, but Baltimore, Washington, or Providence were farther from the main population center and suffered accordingly in the competition.

Physical conditions often closely affect the works of people. On the eastern seaboard human activities, including railroads and highways as well as cities, have focused on key points along the line between plain and hills where carriers must change and where there is maximum access to the largest area.

The environmental influence

It is when one selects a single element of the environment, such as climate, and attempts to base complete explanations on it that the risk of overemphasis is greatest. Climate may be largely responsible for the Eskimo's way of life and for the lack of population in the desert. It is a temptation to use the same simple key to unlock the more complicated puzzle of man's use of the land in more temperate parts of the earth. It would be strange if such an obviously limiting or conditioning factor as climate, one which varies so greatly from place to place, did not have a discernible part in influencing the pattern of human society. But this does not mean that people who live under certain climatic conditions are bound to think or to live in a certain fixed way, or that the climate of some parts of the world gives to their inhabitants a distinctly "progressive" or "backward" character, as many climatic propagandists would have us believe. The environment usually operates as a single system. To isolate one of its factors and ignore the others is to do violence to the actual situation, where there are a host of non-environmental factors operating as well, and where technological change is also taking place. Peoples' own efforts are fully as important as the physical characteristics of the earth on which they erect their civilizations. People standing on a hill are there because the hill supports them, but equally because they have climbed there themselves.[2] The physical environment is important as one of many influences on people acting in concert, not as a determinant. It is both difficult and misleading to examine the environmental influence without continual attention to other influences.

Nature and society in Switzerland

The environmental factor is stressed in the following example in order to make a clear illustration, but the example would not be understandable unless spatial relations and a variety of non-environmental factors were also considered.

Switzerland is for the most part a highly mountainous country close to

Map 8 Switzerland: altitudes and travel routes

the populous lowland area of western Europe (see Map 8). Most of the Swiss population, the cities, and the manufacturing are concentrated on a central plateau, but this area with its famous lakes is surrounded on all sides by high mountains, and there is very little level land even on the plateau. The important basic goods and services produced by the Swiss economy are listed below, and a list of Swiss cultural characteristics follows:

Goods
dairy products and cheese

chocolate
watches and clocks
wood carving
optical goods
lace and embroidery
machine tools
fine textiles
precision machinery
electric generators
electric furnace steel
chemical and pharmaceutical goods

Services

boarding schools
tourism
sanitaria
banking

Cultural characteristics

political neutrality
ancient nationalism
four languages, but no Swiss language
long-standing democracy

Goods

There is a quality common to all of the goods listed: high value per unit of weight and a generally high standard of excellence. Beginning at the top of the list, dairying is frequently practiced in areas which are marginal for field agriculture, usually because of slope, unsuitable or stony soils, or cool, cloudy climate. Such areas may, however, support adequate or even good pasture, and the slope is only a minor disadvantage for animals. Dairying is prominent in Wisconsin, Scotland, New England and New Zealand in part for these reasons. The Swiss mountains offer a more profitable basis for dairying than for field agriculture, and in fact in the high Alps (a word whose original and proper meaning is a high mountain meadow) there are fine pastures. They are free from snow only in summer; hence the general Swiss practice of *transhumance*, or the seasonal movement of animals and people up to the Alps for the summer grazing, from which they come down to lower levels for winter residence. In winter the animals are fed on barn-stored hay, cut the previous summer in the Alps. The Swiss use goats as well as cattle, since goats are better able to subsist on the steep slopes and less luxuriant pasturage, but they have also developed a type of dairy cow, the Brown Swiss, whose exceptional hardiness fits it to its environment without unduly sacrificing the quality of the milk.

The Swiss have the obvious problem of transporting the milk to market, however. Fluid milk is bulky in relation to its value and presents special and expensive transportation problems. The Swiss make cheese, converting the milk into a concentrated and high unit value product which can bear the cost of transportation across mountains. This solution is commonly adopted by dairying areas remote from their markets. Closer in, butter may supplement cheese or take its place, but fluid milk is dominant only within easy reach of the market. This areal pattern is readily observable in Wisconsin in terms of distance from the urban area of Chicago-Milwaukee, or of Minneapolis-St Paul. Switzerland's linear distances to its market are comparable, but they are more expensive distances because of the mountains. The cheese made in Switzerland is of unusually high value per unit of weight; it is very concentrated, as in the case of "Swiss" cheese (the kind with holes in it), or of fancy high quality, as with Gruyère, each piece individually wrapped in foil and packed in attractive little boxes. Switzerland has no particular advantage for dairying or cheese making in comparison with more favored lowland dairying areas nearer to their markets in terms of cost; it must compete on the basis of high quality and the reputation which it wins.

Chocolate making may also be related to the mountains. There are no cacao trees in Switzerland, and the raw chocolate must be imported from the equatorial tropics where it grows. But there is plenty of milk, usually more than can profitably be made into cheese since milk quickly spoils. The Swiss make milk chocolate and convert what otherwise might go to waste into a very high unit value product which has established a world-wide reputation for quality. Milk chocolate can be made more cheaply in Hershey, Pennsylvania. The Swiss transportation bill for inbound raw chocolate and sugar and outbound finished product is much higher; their product must be correspondingly better in the sense that it can bring a higher price.

Swiss watches, perhaps the most famous items of all, emphasize dramatically the solution which this mountain area applies to its transportation problem. During part of the Second World War, when Switzerland was a neutral island in Nazi-dominated Europe, it was allowed to make one monthly shipment to the outside by road as its export quota. The dilemma of what to send was easily solved by filling an automobile with Swiss watches for export; the monetary value of such a shipment strains the imagination, but the cost of haulage was, of course, negligible. Watchmaking also gets around the problems induced by Switzerland's virtual lack of mineral resources. The value of the raw materials used in watches is a tiny fraction of the total finished value, most of which has been added by manufacturing. It pays the Swiss to import the highest quality Swedish steel.

Long experience and skilled labor help to strengthen the Swiss position of leadership in this industry. As their advertisements put it, "Time is the art of the Swiss." Winter is a slack season in a land with heavy snowfall and rigorous temperatures. Relatively little labor is needed to pitch stored hay down to the animals in the barn or to manage the other wintertime aspects of the dairying economy. Something was needed to fill this slack season, profitably if possible. The women made lace and embroidery, and the men did wood carving, as the Nantucket whalers used to carve whalebone and ivory between bursts of activity on their tediously long sea voyages. Little figures, houses, and clocks with wooden works were among the Swiss wood-carving specialties, including the still-famous cuckoo clocks and other precision gadgets, made originally as amusing toys and ultimately for sale. From this kind of beginning came the tradition and skill which go into Swiss watches.

Swiss concentration on high unit value products and on value added by manufacture is clearly carried out in the cases of optical goods, machine tools, fine textiles, precision machinery, pharmaceuticals, and electric furnace steel, all produced in the cities or along the fringe of the central plateau. In the making of generators use is being made of Switzerland's advantage, as well-watered mountain area, for the development of hydroelectric power, but production is also concentrated on high-quality goods. Electric furnace steel is expensive but is of high quality, and in Switzerland it helps to supply the precision metal industries, including watchmaking. Imports of bulky raw materials are minimized, and the emphasis is placed on the value added by manufacture, based on skill, tradition, care, and hard work. Most of the goods which Switzerland produces are related to the fact that Switzerland is a mountainous country near the center of the European market.

Services

In providing services, the Swiss have translated the handicap of the mountains into a profitable asset. Tourism as a major enterprise is relatively young, not much more than a century old on any significant commercial scale. (The extensive religious pilgrimages of medieval Europe and the Islamic world are in rather a different class.) The Swiss, if not the inventors of tourism, were among its earliest modern pioneers. Tourists provided a perfect answer to the transportation problem, especially as average incomes rose in Europe after the eighteenth century and transport facilities improved so that Switzerland was easily accessible and could begin to let tourists pay the cost of developing its transportation network. Tourism became Switzerland's largest single source of income, and the Swiss established a wide reputation as hotel-keepers, makers and demonstrators of sports equipment, guides, and suppliers of fashionable vacations. Other mountain areas have since found such specialization profitable, including Japan, Norway, and northern New England. School pupils and medical patients could be attracted to Switzerland for similar reasons. In education and medicine the Swiss also established a high reputation for performing essential functions with particular skill against the therapeutic or inspiring background of the Alps, whose value finds its way into the school and sanitaria bills.

The force of the environmental influence in Switzerland is clear, but

Switzerland's central location within Europe was also an essential condition for the development which has occurred. Tibet is doubtless more impressive, but few people consider spending their weekends or even summers there. If they went, they would not find the highly skilled specialization of the Swiss in services like medicine, tourism, or hotel keeping, which are themselves reflections of Switzerland's accessibility both to European customers and to the ideas and techniques which are easily spread from all quarters of lowland Europe to crossroads Switzerland. Though mountainous, Switzerland lies athwart important trade routes and is surrounded by several separate major European cultures whose enriching influences have contributed fundamentally to Switzerland's development. The growth of a vigorous, sophisticated, productive culture in any area may often be related to the degree of access and of cross-fertilization or external stimulation which the area enjoys. Switzerland is virtually unique among mountain areas in this respect. The Alps are pierced by several relatively easy passes, and although heavy goods are still at a disadvantage, trade and ideas move freely in large volume and variety. Switzerland separates three major and complementary physical and cultural regions: the Mediterranean Basin, northwest Europe, and central Europe, which exchange with each other to a large extent through Switzerland. Switzerland is what French geographers call a *terre de passage* (land of passage).

Neutrality

The success of banking depends more on Switzerland's political neutrality and on the flow of trade through the country. Neutrality made Switzerland an essential clearing house among the belligerent European powers and gave a greater degree of security to funds held in the country. Switzerland is poor in domestic resources and heavily dependent on foreign trade to make up its deficiencies and maintain its people; trade flows resulting from complementary exchange between north and south Europe provided a further basis for banking and the financing of trade. But neutrality was an important asset. How can it be related to the nature of the Swiss environment?

Neutrality is hard to come by in Europe. Most of the smaller nations would prefer not to be involved in Europe's wars, knowing that they have little to gain and much to lose. But most of them have been unable to resist the military pressures of the Great Powers, and after a futile defense of their neutrality have succumbed in nearly every modern European conflict. Norway, Holland, Denmark, Belgium have all been thus overrun. Only Switzerland (and to a lesser extent Sweden) has been able to hold aloof. A neutral clearing house is useful to belligerents, and Switzerland is conveniently near the middle of Europe. But more importantly, the Alps offer few advantages to an occupying power. Norway, Denmark, Holland, and Belgium all gave to a continental belligerent a foothold on the coast from which the sea lanes could be menaced or dominated or an attack on the continent resisted. Some of the small countries also had heavy industrial establishments of direct military value. In mountainous Switzerland there was nothing, or at least nothing of any use to a belligerent adequate to recompense him for the effort of conquering a country whose natural defenses were so formidable.

Nationalism and language

Mountain cultures are not easily affected by technological and social change, and they occupy an environment fundamentally distinct from the lowland areas around them. The people who lived in the Swiss mountains felt a common bond and strove to distinguish themselves from non-mountain people. Modern Switzerland began as an independent nation-state in the form of three mountain-forest communities called "cantons," which gradually increased their autonomy within the Austrian empire to the point where they could by force of arms complete the separation from Austrian control.

It is in the light of Swiss nationalism and community of interest, related to its mountain environment, that one can best interpret the country's anomalous linguistic situation. There are other bilingual or multilingual states, but with them this difference is a source of friction, as it is for instance in India, Belgium, or Canada. Nowhere has a modern national state approaching Switzerland's small size existed with no language of its own and with four languages in common use. The area of contemporary Switzerland has been successively detached from the surrounding states: Austria, Italy, France, and what is now Germany, hence the origins of its languages. The mountain community of interest was so strong that not only did these linguistic differences offer little obstacle to unity, but there was no need for a single common language to hold the community together. Italian-speaking residents of the Ticino district are as fervently Swiss as the German-speaking residents of the three original cantons, Uri, Schwyz, and Unterwalden, or as the French-speaking people of the Geneva-Lausanne area.

Romansch is an archaic language which has disappeared everywhere else and survives only in a small part of Switzerland. The mountains have sheltered Romansch, given it an area of refuge where it did not rub up against competing influences, and it has been preserved also in part as a badge of regional distinctiveness, much as Welsh and Gaelic have been in their mountain settings. Mountains are nearly always refuge areas for such *relict* cultures or cultural traits, and despite Switzerland's intimate economic, cultural, and political involvement with the rest of Europe, probably on a greater scale than any other mountain area, the remote Romansch-speaking corner of Switzerland remains a backwater in at least this one respect.

Social and political structure

Democracy, in the sense of individualism within a cooperative framework, is understandable as a feature of a hardy pioneering society. However much the mountains may have accentuated such traits, mountains were also involved in a less direct way in two respects of perhaps greater importance. A major basis of the old stratified feudal society of Europe had been ownership of land. There was little good agricultural land in mountainous Switzerland, and there was never a significant split between the aristocracy and the lower orders in Swiss society. Actually, part of the Swiss movement for independence centered on objection to the stratified and aristocrat-dominated society of medieval Austria; nobility and royalty have ever since been banned from Swiss society, except, of course, as foreign tourists with money to spend. Later, as the modern industrial revolution progressed, heavy manufacturing could not economically establish itself in the Alps, and change was concentrated in the precision industries, where skill in the artisan tradition was more important than mass labor. There is still a relatively small spread in incomes among the Swiss population as a whole, and there is little social distinction on economic or occupational grounds. The division between industrial proletariat and industrial capitalists may not prevent the growth of a healthy democracy, but its absence certainly helped to preserve the original egalitarian and democratic nature of the Swiss community.

The Alps make a dramatic environment, and one might expect that the effects on the people who live there are both larger and more readily discernible than in the environmentally more moderate parts of the world, which after all are where most people live. Switzerland is, one may say, halfway between a situation like the north pole, where people have little choice, and the temperate lowlands, where the environment presents a wide range of choice. But the force of the point made by the example of Switzerland is nevertheless applicable to all human relations with the physical world. Although the interrelationship is not often so clear, there are very few situations or developments where a causal connection between people and nature cannot be traced.

NOTES

1. The head of ocean navigation and the geologic fall line may be fairly far apart, and neither may be marked by an actual waterfall.
2. This illustration was often used by the prominent British geographer Sir Halford Mackinder in attempting to explain the dual nature of geographical analysis.

6 THE ECOLOGICAL IMPERATIVE

People have been fighting against the environment ever since they came into existence. What distinguishes humans from other creatures is in fact primarily their ability to manipulate the environment, first (perhaps half a million years ago) by the use of fire, then by progressively more effective hand and mechanical tools, and finally by a great variety of machines and chemicals. The last hundred years have seen this process accelerate tremendously, as people rapidly increased their abilities to manipulate the environment through industrialization. They have probably made more destructive impact on the environment during the past century than during the whole preceding period of human history, an impact which is still continuing in full force. The change is a revolutionary one, and one which has only recently and belatedly been recognized as posing a serious threat to human survival itself.

In the simplest terms, *ecology* is the study of the mutual interplay among and between organisms and their environment; originally a biological concept only, it has recently been expanded to include, and now to emphasize, the role of the human organism and human inter-relationship with those parts of the natural environment which are important to them. In this context, the nature of the environment as a set of interlocking systems in balance with one another is especially clear, a balance easily disrupted by aggressive human powers, the part of the ecological system which has grown out of control.

The natural environment is a fragile construct, or at least it is danger-ously vulnerable to the destructive manipulation which people have practiced on it. Other species have adapted to their environment or perished. People survived a major environmental change in the form of the last glaciation with the help of new weapons, of which fire is an instance, but their newer weapons have come to threaten the very base which sustains them. On a human time scale, the environment changes very slowly as a result of its own processes. People have made themselves in effect an external force acting on it, in such a way as to produce rapid and sweeping changes. They have divorced themselves from their environment and as a result may have cut the cord which nourished them. Without this new external pressure, the environment would in time heal itself, but the pressure of modern human technology is unrelenting and still increasing. It continues to disrupt the comparatively slower and more fragile processes by which the various elements of the environment remain in a dynamic balance.

Awareness of the dangers involved in disrupting the natural processes is very recent, except for a few far-sighted individuals who, in the early stages of the industrial revolution, realized some of its destructive consequences and warned against them. *Like most prescient people in any age, they were ignored*. Their warnings were easily put down in the atmosphere of enormous confidence in man's supposedly unlimited powers, brought, understandably, by industrialization. The Victorian and post-Victorian West were supremely confident that Western man had created a new kind of "progress," and through his new industrial technology of steam and steel had finally "conquered" nature. That confidence has begun to be undermined only in the last few years, and it takes a great deal of time to slow down, let alone reverse, almost any aspect of mass behavior, especially one which has been supported with the degree of enthusiasm given to industrial technology and its fruits. Nor can it be denied that industrial technology has, at least in the short run, tended to improve the material basis of life for most people in the areas where it has spread, including its effects on public health and food production.

The longer-run effects of the internal combustion engine, the use of chemicals in agriculture, food processing, and industrial products, nuclear weaponry or testing, and the burning of fossil fuels (coal, oil, gas) may, however, tip the balance sharply to the negative or even the fatal side. Awareness of these risks is still so recent that we really do not yet know their dimensions, nor how best to guard against them, short of abandoning the whole structure of industrial technology. Such a solution is of course unrealistic even if it were appropriate, but each year it becomes clearer that a drastic change of some sort in human behavior must take place, and as rapidly as possible, if we are to avoid altering our environment to the point where we can no longer survive in it. We cannot of course eliminate it; we can, however, eliminate, or disrupt or destroy those parts of it which sustain us.

Attitudes and awareness

Public concern over this problem, however belated, is encouraging, for it is only when large numbers of people become excited about an issue that mass behavior and public policy are likely to change. But unthinking commitment to what appear to be the most dangerous aspects of industrial technology has been affected so far only to a minor degree. These things include the unbridled use of the private car, nuclear explosions, defoliants, plastic containers, poison sprays, and food additives. Most people are not yet at the point where they are willing to give up any significant personal convenience, or established habit patterns, in order to slow down what may prove to be disastrous or even irreversible destruction of the environmental base for future human life. Most people are notoriously short-sighted.

It is reassuring, however, that a number of academic disciplines have turned their attention increasingly to the vital interrelationship between people and their environment and to the potentially or actually disastrous consequences of many aspects of human behavior in this connection. Anthropologists in particular have been concerned for some time with the ecological nexus and with nature-culture interactions in their regional variety. This concern has been taken up by more and more anthropologists in the past few years and seems likely to present an enduring (if overdue) change in the discipline. Historians, sociologists, political scientists, and even a few economists have also picked up this theme, and one may hope that they will persist with it.

The natural sciences have always been concerned of necessity with ecological analysis, and indeed their work has been primarily responsible for the existing body of knowledge, but in recent years they too have given increasingly broader attention to the outstanding human dimension of the problem. Geographers cannot afford to look down their noses at others, however, since although in principle they have always had a commitment to environmental study, the overwhelming bulk of geographical work between about 1940 and the end of the 1960s represented anything but such a commitment; studies of the human-environment interconnection are still only a fraction of all geographical work, although they are rapidly increasing. Some of this reluctance to address what would seem to be a central concern of the discipline doubtless stems from the earlier reaction against the extremes of environmental determinism. But it is now no longer out of fashion for geographers to consider, as the nature of their discipline requires, the interrelation between people and their natural world. In an age of galloping industrial technology and headlong global population growth, there can be no more important problem for study. It has taken nearly two hundred years since the beginning of the industrial revolution for overconfidence to begin to give way to the double realization that both people themselves and the natural environment they have learned so effectively to exploit are terribly fragile. The reason of course is that they are truly interdependent. As we "conquer" or weaken the ecological context in which we exist, we weaken or destroy ourselves. In the longer run, people may prove to be more fragile or vulnerable than nature. *It can exist without them; they cannot exist without it.*

Ecological imbalances

People have been hacking away at the forests with ever-increasing effectiveness since early Neolithic times, and in most areas have then introduced agriculture. (Some of the dimensions and consequences of deforestation are examined in Chapter 9.) They have replaced one vegetational system, the product of many millennia of adjustment to climate, soils, water resources, fauna, and other elements acting in concert, with another which is a largely artificial construct. Most agricultural systems concentrate on the growing of plants originally derived from other and distant areas and refined by selective breeding over time so that they little resemble the original wild parent; most of them cannot maintain themselves in competition with wild species. Natural landscapes are occupied by a series of plant communities, a variety of species sharing a given environment in balance with one another and with the local fauna, all in adjustment to the other environmental characteristics of the area.

Agriculture removes all this and substitutes for it a single plant species, at least in any one field. Some farming systems may grow several crops (although rarely in the same field), but others concentrate heavily or exclusively on a single crop, such as cotton, rice, corn, or wheat. The greater the departure from the original varied complex of plant communities, the greater the risk that the ecological balance may be seriously disrupted. Each plant has a different set of requirements for soil, moisture, temperature, and sunshine. To cover an area with a single species instead of the mix which originally occupied it is to impose unnaturally heavy demands, which usually have to be met in part by artifical means: irrigation, fertilization, and the elimination of competition through weeding. In the longer run, however, irrigation and fertilization cannot sustain productivity or soil qualities indefinitely. In an agricultural system, the plants and often also their residues are removed from the field instead of being allowed to complete the natural cycle of growth and decay, returning the decomposed materials to the soil and preserving both its fertility and its structure. This cannot adequately be compensated for with the use of chemical fertilizers, especially when the soil is supporting only a single plant species. Over a period of time irrigation tends to concentrate in the soil an accumulation of minerals carried in solution or may also lead to waterlogging. In arid and semi-arid areas, in particular where irrigation is most needed to sustain agriculture (and where it is therefore especially unnatural), the accumulation of minerals or waterlogging may relatively quickly lead to declining yields and eventually may make the soil wholly unproductive, in the absence of adequate rainfall to flush it or of natural drainage to keep it free.

Concentration on a single plant (or animal) species also greatly increases vulnerability to disease. Natural plant and animal communities are provided in effect with a series of fire breaks; most diseases are specific to particular species and cannot spread through other species, but multiply very rapidly in an area covered by a single crop plant. The same is true of most insect pests. Poison sprays used to attempt to control such ravages have already been shown to present risks to human life which may be worse than the problems they are designed to prevent. Heavy reliance on chemical fertilizers to maintain yields may also adversely affect the nutritional qualities of the crop. This discussion should not lead to the hopeless conclusion that deforestation and agriculture were all a terrible mistake and that we should return to a hunting-and-gathering culture. But it does make clear that alteration of the environment can be a two-edged sword and that it must be done with care and with some knowledge of the possible consequences. With careful management and an awareness of the fragility of the ecological balance, the worst mistakes can be avoided and the damage minimized.

The most destructive, and hence most fragile, forms of agriculture are those which concentrate exclusively on a single crop, which is grown year after year. Cotton is a good (or bad) example, especially in terms of the history of many cotton-growing areas in the southeastern United States, whose soils were exhausted and eroded to the point where agriculture in any form was no longer possible in large parts of many formerly productive cotton regions. The area's demise was also hastened by the disastrous spread of cotton diseases and insect pests, including the boll weevil, which found this artificially uniform environment an ideal hothouse. (Fortunately, irrigated rice, globally the most important crop grown on a *monoculture* basis, is less susceptible to disease, and paddies are protected from erosion by diking, terracing, and their cover of irrigation water, which also brings in plant nutrients, so that with the addition of careful fertilization with organic wastes, yields can be maintained over very long periods.) A better and more stable ecological balance can usually be achieved through the use of crop rotation, by growing several crops (in different fields) simultaneously and by combining annual field crops with tree crops and with animal husbandry, thereby also providing a source of manure. However, such a solution reduces rather than eliminates the problems which any agricultural system creates by constructing what is necessarily an artificial environment and hence disrupting the ecological balance. Any alteration of the environment carries the risk of consequences which may be more detrimental than beneficial, in the short or long run.

Artificial introduction of animal species may also have unforeseen and harmful consequences. The introduction of the rabbit into Australia is probably the most dramatic case, since in the absence of natural enemies it multiplied out of hand and severely damaged the environment for most other species (including people). Involuntary introductions, such as the spread of the Japanese beetle into North America in the 1930s, have been only slightly less destructive; the starling, referred to in Chapter 4, has become a major pest. Both species found in effect an ecological niche in North America not occupied by other species, and became successful enough to upset the balance so that they made severe inroads on other plant and animal life, including for the most part species and plants which were considered far more valuable or desirable. The use of imported species to fight what are regarded as pests is always dangerous because the results are unpredictable and may backfire. Hawaiian sugar-cane growers some years ago introduced the Indian mongoose, hoping that it would eliminate or reduce the rat population which had mushroomed in the cane fields and was doing great damage. Instead, the mongoose coexisted more or less peacefully with the rats and became an almost equally important pest to the cane growers.

Sewers in the sky and the sea

Agriculture at least attempts some degree of cooperation with and awareness of environmental conditions. Other human activities, especially during the past century or so, represent a far more drastic departure from the concept of ecological balance and have for the most part been undertaken with no regard at all for the possible consequences, or in complete ignorance of them. The pollution of the atmosphere and of water bodies has reached nightmare proportions as a result of industrialization and the attendant explosive increase in the burning of fossil fuels, the massive use of chemicals in manufacturing, nuclear bombs, defoliants, and the mushrooming of the internal combustion engine. Most people have become aware of these problems, but the solutions require more drastic change in human life styles than most people seem ready to adopt. Although we still know distressingly little about the dimensions, trends, and consequences of atmospheric and water pollution, we know enough to justify profound alarm.

FREQUENCY OF INVERSIONS AT LOW ALTITUDES—FALL SEASON

Map 9 Temperature inversions in the United States

Acute smog builds up under certain atmospheric conditions of temperature, sunlight, and air movement (or its absence). Sunlight in particular acts on the gases emitted from automobile exhausts and certain other industrial emissions to produce ozone, a toxic gas. In the absence of air movement, or when what is called a temperature inversion exists (a layer of colder air trapped below a layer of warmer air so that the whole system is static instead of rising), toxic concentrations build up instead of being dissipated (see Map 9). This is the common pattern in the Los Angeles area, which lies in a coastal basin surrounded by mountains, and in other parts of the California littoral where warm temperatures, abundant sunshine, cool air off the coast above a cold ocean current, and massive car and industrial emissions in relatively enclosed basins combine to produce both chronic and acute smog. Certain plant species, especially such cultivated forms as lettuce, may show damage first, but effects on the human population may be equally great, at least in the long run, and far more serious than the eye-watering and coughing which are short-run symptoms of irritation.

Conditions conducive to acute smog are fortunately not continuous; temperature inversions are more uncommon than common even in the Los Angeles Basin, and there is usually some air movement. Again, people have for the most part perceived or adjusted to the usual rather than to the unusual, as they have done in other areas where acute smog is a recurrent and even lethal phenomenon, mainly in other enclosed or semi-enclosed basins where there is a low, dense cloud layer and a high concentration of industrial, household, and automobile emissions. The London Basin is a famous example, but to a degree also an encouraging one in the sense that the smog hazard was perceived and counter measures taken to restrict emissions. Most of the London urban area was declared a "smoke-free" zone, and although the law was not applied literally, emissions were greatly reduced and acute smog became less frequent. Unfortunately, it had to kill or cripple a great many people before the hazard was seen clearly and urgently enough to produce some action. Unfortunately also, the common response to dramatic disaster is temporary; chronic problems are not so easily perceived nor is behavior so easily attuned to their risk.

Systematic study of this problem has begun at the eleventh hour, when it may already be too late. Atmospheric pollution affects a huge and complex system of air circulation and regeneration. We have been pouring gases and particulates into the air at an accelerating rate, especially since about 1900, and can be certain only that the effects are bad enough to warrant severe control measures. We do not yet know in any detail how and to what extent atmospheric pollution may affect weather and climate on a macro scale, nor the full scale of its effects either on plants and animals or on humans, except that whatever affects one part of the ecological system affects people, who are an interdependent part of it, whether or not it directly gives them silicosis, emphysema, or cancer. The oceans also constitute a vast system of circulation and a series of ecological systems with which people are involved as users of oceanic resources. It has been startling to discover in recent years the extent to which even the major ocean basins have been polluted and altered by human action, a combination of haphazard waste disposal and over-fishing which has already changed fundamentally the entire marine system and which may have consequences nearly as serious as those of atmosphetic pollution. Although we still know relatively little, mainly because study has begun so late, it is clear enough that pollution control presents no insuperable technological problems. The means are straightforward and technically well within our power. The problem is to persuade people, corporations, and governments to apply these means and to accept the higher costs and the inconveniences which would probably result in most cases. Most people find it difficult to make short-term sacrifices in order to maximize their long-term welfare. But in ecological terms, the hour is late and drastic measures are called for.

Change and process

The burden of human impact on the natural environment has been greatly increased not only by industrialization and the revolution in power use but by the rise in human population which has resulted from the increase in productivity and longevity. This problem is discussed in greater detail in Chapter 13, but it needs to be mentioned here as perhaps the most basic aspect of the ecological revolution. We have upset the ecological balance primarily by multiplying our own numbers and by spreading our occupance and use of the environment over almost the whole of the world, while at the same time increasing the variety of our demands and impact on the ecological system. Although increased human survival and greater numbers may be or may seem a strongly positive matter, it clearly has negative consequences as it further strains or subverts a total system on which people are just as dependent as are its other components, especially as we continue to weaken those other components. Finally, one may raise the more philosophical question of whether people can be preserved at all unless we can learn not to destroy ourselves, indirectly by destroying our environment or directly through war and hate. People, the supposed conquerors of nature, have much to learn. Perhaps it is the beginning of wisdom that we have at last come to recognize some of our own limitations.

Fortunately, neither people nor their environment are static. The ecological system is best seen as *process* rather than as a fixed set. As part of this system, people change too and will continue to change, hopefully more responsively to their ecological setting than heretofore. The environment as a whole, and each part of it, is continually changing and adapting in a mutually linked system. Diurnal, monthly, and seasonal alterations of this sort are obvious, but on a much larger time scale the environment can also be seen as dynamic. Soils and vegetational complexes, for example, alter slowly over time in a mutually responsive fashion, and also in response to changes in climate. It is possible to reconstruct these changes throughout most of the earth's history by the examination of fossilized evidence of fauna, flora, and major geologic epochs, and to measure their magnitude. Humans appeared on the scene extremely recently on such a time scale and their long-run place in the ecological system is therefore still unclear; no doubt much will depend on how well they are able to adapt.

Analysis of plant pollens preserved in the soil is a technique, developed particularly in the past thirty years, which makes it possible to pinpoint with considerable accuracy even short-term and minor changes in vegetation, and hence in climate during this brief period in which humans have been present. Pollen grains are extremely well protected against decomposition and survive in the soil under most conditions for twenty or more millennia. This power of survival extends their evidence in time well into the period of the last glaciation. As pollen from plant species which tolerate cold, drought, dampness, or heat vary in their occurrence and mix in a given area over time (temporally sorted by their vertical stratification), one can see climatic changes mirrored over periods as short as a few centuries. During the evolution of the human species, people have had to adapt to drastic changes in climate and in other elements of the environment. The emergence of *Homo sapiens*, on the order of fifty thousand years ago, coincided with the last advance of the ice, and people began to evolve toward agriculture, domestication of animals, and ultimately civilization during the equally drastic changes accompanying the subsequent retreat of the ice sheets and the return of warmer climate. Humans have so far thus proved themselves eminently adjustable and taken maximum advantage from both major and minor environmental fluctuations rather than being limited or eliminated by them, as has been the case with many other species. The greatest challenge, however, is of their making: having reduced the one-way power of the environment over them, can they now continue to manipulate it without destroying its ability to sustain them? Can they in other words, change now to adapt to the new problems which they have themselves created by upsetting the ecological balance? Or will they be eliminated, as so many other species have been, by their inability to change, to reverse the present disastrous course?

Much of the answer will depend on how people perceive the environment: as obstacle, as enemy, as tool, or as the basis of their existence which requires nurturing care. Culture forms a lens through which people perceive reality. Since culture changes over time, and is different in different places at the same time, the distortion of this lens is not constant. It is perfectly possible, in other words, for people to alter their perception of the environment in the face of this new challenge so as to adapt their behavior accordingly. We may already be seeing the beginnings of such a change, as people in a variety of contemporary cultures become more sharply aware of what are in fact relatively long-standing problems such as atmospheric pollution, to appreciate their urgency, and to be ready to do something about them. General standards of what constitutes good or acceptable or clean air have changed radically and rapidly since 1900, since 1950, and since 1970 in this country as the atmosphere has been progressively degraded and the problem has consequently forced itself onto the attention of more and more people. The same phenomenon has been repeated in western Europe, the Soviet Union, and Japan as industrialization and the spread of the internal combustion engine have reached similar scale. Tokyo, for example, which by 1970 probably had the most polluted air of any big city in the world, may also now be the most pollution-conscious. In other words its citizens may perceive atmospheric pollution especially sharply. At major street junctions in the downtown area automobile pollution sensors have been mounted which flash continuous readings of the changing count on huge illuminated billboards. This was done as part of a carefully planned campaign to win public support for rigorous pollution controls, especially on engine exhausts and industrial discharges. Pressure, including new laws, was put on car makers and factory owners; after the expected complaints, they complied, and the controls proved quickly effective. Tokyo air and surface water bodies became the cleanest among the world's big cities.

Perception of hazards

Even in Japan, Tokyo and other cities continue to grow, and private car ownership and use continue to multiply. Some groups object to the cost of pollution controls, however shortsightedly, conveniently forgetting the deaths clearly attributable to pollution before it was reduced. The situation is comparable to human behavior in response to other environmental hazards, even those which are clearly perceived. Although the environment as a whole changes relatively slowly over time, periodic extremes may occur every few years, such as drought, flood, earthquake, volcanic eruption, and (one must now add) concentrated as opposed to chronic smog. These and other such dramatic events are well recorded and reported. They can be charted over time and hence to some extent predicted, at least in the sense that they can confidently be expected to recur within a certain period, and in certain well-defined areas. In general, people tend to disregard this by adjusting their behavior to the usual rather than to the unusual. Settlement has not by any means avoided the San Andreas fault area in California despite the disastrous earthquake of 1906 and the frequent reminders that another major earthquake will occur there. The slopes of Mt Vesuvius are settled and farmed, despite recent volcanic activity. The explosion of Mt St Helens in Washington state in 1980 was a dramatic reminder of the enormous force of vulcanism, and the death toll underlined the need to calculate the risks involved. About an eighth of the United States population lives in areas subject to periodic flooding; a similar proportion lives in areas subject to periods of major drought, although drought is more difficult to define and may not so immediately affect the welfare of non-farmers. At least one-third of the United States population lives in areas of chronic and periodically acute smog. Perception of environmental hazard is obscured to some degree, although as pointed out above it is changing. In general, people continue to pay more attention to what they perceive as the rewards of living in environmentally hazardous areas than they do to the risks, and they discount the unusual, suppressing or perceiving only vaguely the certainty that the unusual will happen again, and probably many times in their lifetime, since it is the nature of the environment to fluctuate.

Human survival?

Fortunately, perceptions and behavior change. In a world where there is beginning to be nearly global exchange of information and awareness of differences in experience and culture, each area may also be able to benefit from the perspective, insights, and mistakes of others. All people share a planet whose environment operates as a single system. What people do to it in one area is likely to affect its nature elsewhere. Nuclear testing in the south Pacific or central Asia puts radioactive particles into the system of global atmospheric circulation and into the similar circulation system of the oceans so that all parts of the world are covered. Industrial pollution in any area also enters the global circulation system. The effects of biological and chemical warfare in Vietnam cannot be confined to that unhappy area. We are all in the same boat, even if some of the most obvious leaks are only in certain parts of it. The more industrialized nations must take the biggest share of the blame, partly because they are the chief polluters and also the chief users of finite resources, partly because their degrading effects on the environment are inherited by everyone else, the poor two-thirds of mankind, whose economic margin is much smaller and who therefore can less well tolerate the consequences of a deteriorating environmental base. The United States alone, with less than 5 per cent of world population, accounts for over one-third of world consumption of fossil fuels and generates nearly half of the world's mechanical energy, producing in the process nearly half of the world's atmospheric and water pollution. This is both irresponsible and short-sighted and will be increasingly resented. Change is urgently overdue. All mankind is interdependent, and in turn is interdependent with the environment.

Something may be learned from the experience of other cultures, if only the realization that perception of the environment and attitudes toward it have varied greatly from place to place and from period to period. Most of traditional Asia, for example, saw people as an integral part of a cosmos dominated by nature. Contentment as well as material success could come only through acceptance of the rightness of people adjusting themselves to the greater natural world of which they were a part. This attitude was to some degree common to many agrarian societies elsewhere, but in China especially it was enshrined as a central part of a philosophical and moral system. The bulk of imperial administration was devoted to the care of the land. The elite and the peasants alike evidenced a deep respect, almost a reverence, for a natural order conceived as grander than humans and more to be admired. The Chinese certainly altered nature, through terracing, irrigation, deforestation, and intensive cultivation of an immense landscape. But such activities were not seen as pitting people against nature but as carrying out the role of careful and respectful stewards. Cooperation with nature was necessary; it could benefit people only if they accepted the limits it was seen to impose. With this kind of care, the Chinese environment became without question the most productive agricultural area in the world, and by far the most consistent, continuing to produce high yields for over two thousand years.

The Western attitude toward nature, though more varied, has tended instead to see nature as an antagonist to be conquered. Even long before the industrial revolution the West enthroned humans as God's supreme handiwork, and tended to assume that they had in fact a duty to subdue nature. Only by doing so could they achieve the "progress" in which the West also believed. Where in the traditional Chinese view mining was devalued and even considered impious because it robbed the earth instead of making it more productive (as agriculture was seen to do), Western thinking emphasized the human mission as being to wrest nature's secrets from the earth and to exploit it. As industrialization progressed, metals and minerals were seen, through the machines they made possible, as "freeing" people from their earlier dependence on nature and even as cancelling out the importance of the environment as a factor in human affairs. With the advent of industrialization in Asia, and of an originally Western ideology (Marxism-Leninism) in China, many of these Western attitudes toward the environment have spread eastward. To a great extent, of course, industrialization is not culture-bound and brings about a fundamental change in attitudes and behavior wherever it spreads and whatever its cultural origins or adopted context. But some of the traditional Asian attitudes remain, even in highly industrialized and urbanized Japan or in Communist China and developing India. Perhaps Asians will be able, with the help of their traditional values and perceptions, to make a less disastrous accommodation between industrialization and the natural environment than the West, with its different traditions and perceptions, has managed to do.

In Japan and China especially there are already some signs that this accommodation may be beginning to happen in the form of controls and in the form of regional planning for dispersal. Technologically also we still have much to learn about the consequences of and the means for controlling environmental degradation. New technology is continually being evolved; the directions it takes are determined in large part by felt needs:

what a given society considers important or urgent, which is in turn dependent on that society's perceptions. In parts of Asia there may well be a sharper perception of the environment and of the urgent need to limit its degradation than elsewhere in the world. Having pioneered industrialization, the West may well now have to learn from the East how to make industrialization viable and livable. One of the salient characteristics of Asian society has always been its emphasis on the general welfare as opposed to individual license. This emphasis has been most consistently and strongly manifest in the family system, but it has also shaped behavior and policy in the large. Contemporary China is a clear example, but the tradition of individual subservience to public good is in fact very old in China and in most other Asian cultures.

The West, especially the United States, has tended instead to emphasize individual freedom – "rugged individualism." In the frontier past, while space and economic opportunity were ample, such an attitude was understandable and viable. In a more crowded world where individual actions profoundly affect public welfare, they may be less so. In any case, environmental degradation can be checked only by concerted action, something for which there is old and varied precedent in many Asian societies. Perhaps they can lead the way or perhaps the West can alter its own perceptions in its own way and can begin before it is genuinely too late to reverse the disastrous course which industrialization has set. We live in a closed ecological system; it can nourish or extinguish us; the choice is ours.

Nature is full of enormous vitality. We need to nurture and enjoy it. The environment can purge and heal itself if we will give it a chance. But first we must learn to understand it, to respect it, and to interact with it positively rather than negatively. Instead of destroying it, we need to learn how to let it sustain us.

7 CULTURE, NATURE, AND CLIMATE

It is difficult to consider adequately any aspect of human settlement on or use of the earth without considering the environmental influence. But it is perhaps best to think of the environment as a *permissive* rather than as a causative factor, in terms of the range of choice or the kind and degree of opportunity which it offers. In a harsh setting like the polar ice-cap, people have little choice. Without a highly developed technology imported from some other area, they can maintain themselves only in a strictly circumscribed manner, and there is little opportunity to specialize or to branch out into other activities. In harsh environments human settlement is usually characterized by a low standard of living and a primitive level of development. The business of staying alive is demanding enough to leave no room for the accumulation of a surplus or for the development which depends on a surplus.

Our dependence on nature

It is possible to argue that we become more rather than less dependent on the physical environment as our technology, economy, and society become more highly developed. Although we may be able to irrigate the desert, heat houses, fertilize the soil, chop down the forests, and tunnel under the mountains, we do not remove the limiting conditions which these phenomena imposed upon us but merely exchange one set of conditions for another. It costs money to "conquer the forces of nature," usually a great deal of money. It pays to spend this much money only under certain circumstances, part of which are usually a reflection of the physical environment. Tunnels pay only with an existing or potential high volume of traffic, irrigation pays only for a present or potential highly productive agricultural system, and so on. Each of these is in turn depen-

dent on what the physical base and its relations in space will allow. Once
the technical improvement has been made, human choice is also limited.
The return from such a dearly bought piece of property must be maximized, and therefore it can be used only for certain restricted high-value
or highly productive goods or services. Each such technological alteration
of the environment usually requires a more or less complex system of
organization to create the alteration, to maintain it, and to operate the production system dependent on it. Changes in the environment which new
technology enables people to make may also create new resources which
fundamentally affect human society. On the other hand, people may
destroy resources or alter the environment in such a way that human
activities suffer or are limited.

Before people had these technical abilities, they were, of course,
obliged to walk around or over the mountain, or avoid the desert unless
they chose to be nomads. But the mountain and the desert really meant
less to them, in terms of effort or of dollars or of the distribution of routes
and settlements or of economic production, than they do now. As
specialization proceeds, interregional exchange also becomes increasingly important. Each region is concerned with the economic, political,
and cultural complex of many other regions, some of which may be
remote but complementary. Each region is thereby concerned, too, with
the environmental basis of other regions and with the aids or barriers to
movement across the intervening space. The weather in the Australian
wheat belt is important in the markets of Liverpool and Chicago and on
the farms of the Dakotas; the fact that these scattered places are connected by the cheap volume transport created by the industrial revolution[1] is responsible for their mutual sensitivity to each other's
environment. The transformation of tropical rain forests into rubber
plantations has made the local environment far more important to people
than before the forest was altered.

Specialization also means that we are dependent on many more resources
than we used to be. To the American Indians of the fifteenth century, iron
ore in Minnesota or coal in Pennsylvania meant nothing. They had no
worries about the remoteness of the ore, the cost of its shipment to
population centers, the freezing of the Great Lakes, or the cost of mining
and shipping coal from the Pennsylvania mountains. If their wells or
streams ran dry, they could always move and did so frequently as a matter
of course. A steel mill is much more tightly bound to its site and to the
physical conditions existing there to which it is extremely sensitive,
especially the supply of tremendous quantities of fresh water for cooling
and flushing purposes.

Each economic specialization which we develop makes us more sensitive to both local and distant environments, because it increases our
dependence on specific or varied resources and our dependence on
exchange. A modern fruit orchard, for instance, contrasts markedly with
the Indian cornfield or hunting ground which might have occupied the
same site three hundred years ago. The fruit farmer must be concerned

with many local conditions which were more nearly matters of indifference to the Indian – early frosts, air drainage, insects, hailstorms, and other things in which the environment is partly reflected, such as the price of land and the availability of low-cost labor at picking time. He is also concerned with the same conditions in other fruit-growing areas with which he competes in the national market. The price of New York State apples (and hence the fruit grower's profit) will depend in large part on the quantity and quality of the Washington State apple harvest and the price tag attached to it; it will also depend on the condition of overseas markets and on the competition from other fruits, or simply from other food items, for a share of the housewife's food budget. Each of these will in differing degrees reflect environmental factors.

There is, of course, one vital respect in which people have gained the upper hand over their environment. They may still be limited by the environment in their economic decisions and developments, but by and large they no longer freeze or starve to death in proportionately large numbers, as they did until about a century ago, because nature has been niggardly or vacillating. Exchange evens out most regional and temporal variations in production at the same time that it makes people more sensitive to the environment. Perhaps this is progress enough to say that we have conquered nature, but it is a limited conquest which has done the reverse of eliminating the environment as a factor in human affairs.

What is currently referred to as "the energy crisis" is probably the sharpest single reminder of the relevance of the environment to human affairs, although the problems it encompasses have been with us for at least the past century. The harnessing of energy in a variety of forms in the course of the industrial revolution has been the single most important factor in raising living standards world-wide. This has happened through the development of power-driven manufacturing of an ever-growing range of goods in rapidly increasing quantities beyond what was possible by hand, and at lower cost. New sources of energy, including chemical fertilizers (made industrially from coal, oil, and other chemicals), power-driven pumps and dams for irrigation, and farm machinery powered by fossil fuels have greatly increased both agricultural yields and total output. Finally, power-driven transport by land and sea has made it possible to move large quantities of bulky goods, including food and the fertilizer and machinery to grow it, plus other basic commodities over long distances at bearable cost. Only occasionally in isolated areas, such as the Sahel of northern Africa in a series of drought years in the 1970s, do large proportions of given populations now die for lack of food. Such famines, and related problems of malnutrition, health, and inadequate housing or clothing are mainly economic, social, and political in origin rather than resulting from true physical shortages or physical inability to distribute surpluses from elsewhere. This does not make such problems excusable, but they do not stem from weakness of human control over energy to correct them.

The enormous increase in human use of energy, derived mainly from fossil fuels (coal and oil) over the past century has risen especially sharply in the last few decades. The amount of energy use per capita is in fact one of the basic measures of economic growth and development, as it enables or contributes to increased production of all goods and services, including many not available before. Everything we produce, transport, or consume, even services such as medical care or education, depends on the supply of energy. To this extent, we have become more dependent on the environment rather than less, since the environment is the source of all energy. So far the predominant forms of energy have been developed from finite and non-renewable materials, mainly coal, oil, and natural gas. Present use rates will exhaust known supplies of oil and gas in the short-run future, and will at the same time continue to push up their costs as a result of scarcity and the rising costs of extraction from off-shore (under water) fields, deeper pools, or oil shales.

Such factors have already made a fundamental impact on nearly all aspects of most national economies, as surely every citizen by this time is personally aware. Coal is more plentiful, but still finite, and its costs will also go on rising. At least equally if not more important however is the environmental impact of continued dependence on fossil fuels, especially coal, whose use has already seriously damaged the environment through atmospheric and water pollution to an extent which clearly threatens human welfare, as has the combustion of oil, industrially and automotively. As pointed out in Chapter 6, one of the most worrying aspects of this problem is that we do not clearly understand how bad or how irreversible this damage is or what its longer-term effects may be, on human life and on the rest of the environment which sustains us. Even more disturbing risks are associated with the development of nuclear energy.

Our exploitation of the environment as a source of energy has perhaps created as many problems as it has solved, and has in any case dramatically underlined our vital dependence on what the environment provides for our support. It has also underlined the environment's fragility when subjected to the pressures and abuses which industrialization and population growth have brought to bear on it. The harnessing of energy has enabled a huge and rapid increase in world population, as the means of subsistence have multiplied, but this in turn has put even larger strains on supplies of energy, and on the environment as a whole. We can escape from these dilemmas only as we succeed in developing renewable and non-polluting sources of energy, such as solar power, and as we can reduce our present profligate use of energy. This and other aspects of the energy problem are discussed in greater detail in Chapter 14. For the present, it needs to be emphasized again that we are as interdependent with our environment as ever, and that as human technological powers have increased we have merely changed the nature of that interdependence, but without in any way freeing ourselves from environmental constraints.

Each culture and each period of human development may be regarded as a different type or a different stage of adjustment to a given environment. Although we may make this adjustment more productively as our technology improves, the need for adjustment is not thereby disposed of and may indeed become more demanding as a greater variety of environmental factors or resources become important. It should also be clear that the process of regional specialization which technological improvements and cheap exchange make possible continues to create greater and greater regional diversity over the world, at least in terms of production, rather than making regions more alike. This is so despite the degree of uniformity which improvements in technology and exchange may also spread in many other respects. We are no longer all farmers, nor do all agricultural regions produce the same kinds of things or attempt to produce everything they need. Though industrial cities may soon be world-wide in their distribution, and with them the other trappings of an industrial-commercial civilization, each larger region and each subregion within it grows more rather than less distinctive as it becomes more specialized or as it is increasingly able to make special use of its total set of environmental and other advantages. Technological change thus leads in this sense to an increase rather than a lessening of environmental influences on people, as it also leads to an increase in spatial interaction, which is in part an expression of influences between different environmental bases. Location is important only if there is interaction, if there are other places related to the place in question. As these relationships increase, so do the influence of location and the influence of varied environments increase in their impact on human activities.

The environmental complex

There are five basic elements of the physical environment which tend to operate as a unit: climate, vegetation, soils, land forms, and water resources. Usually in any given area they tend to be consistent with each other. A certain climate will tend to be accompanied by a characteristic set of vegetation, soils, land forms, and water resources. Mineral resources other than water are not related to these five in their distribution or are related in less current or less obvious ways. They do not occur consistently or directly as part of the same physical complex to which the five basic elements belong. Current climate, for example, has no consistent relationship to coal deposits. Ores and other mineral deposits are certainly part of the physical environment, but whether they are usable resources in a particular case depends as much on economic, locational, and cultural considerations as on the physical facts. Many mineral deposits which are loosely called "resources" are not so at all, because they are not economically usable; the word *resources* could be better applied to climate or soils or to human skills. The meaning of resources as an important part of the physical environment will be discussed in detail in Chapter 13. For the present the discussion will consider the physical environment in terms of the five elements of climate: vegetation, soils, land forms, and water resources.

WORLD POPULATION DISTRIBUTION

Population Density, per square kilometer

Under 1

1 to 25

25 to 100

Over 100

SCALE

GOODE HOMOLOSINE EQUAL-AREA PROJECTION

Source: U.N. Demographic Yearbook
1980

Map 10 World population density. *Based on data for the late 1970s, slightly different years for different countries.*

The major role of climate

Climate is by far the most important aspect of the environment; most of the other environmental factors are in part reflections of it. Few human developments can ignore climate or be independent of it, and climate's permissive or conditioning influences are apparent far beyond the obvious sphere of agriculture. Climate more than anything else about the environment sets the limits of what people can physically or economically do. It offers or denies them opportunity or special advantages and is therefore directly involved in the distribution of population over the earth and the particular economies and societies which have arisen in different regions. The great empty spaces of the world, from a human point of view, are predominantly the work of a harsh climate, and the great clusters of population are generally coincident with areas where the climate is broadly permissive. Within the range of choice which a permissive-type climate offers, people may develop much or little, in accordance with a number of non-climatic factors, some of them pertaining to other aspects of the local environment, some to relations in space, and some to matters such as cultural tradition, political management, or social structure. But climate may be said to set the stage on which people may play their role by the manipulation of other less uncompromising factors.

Human control over climate

It is still reasonable to assume that we can do very little to change climate. We can heat houses or install air conditioning, carry an umbrella or wear warm clothing and thus be saved some physical discomfort. But we do not thereby escape the economic limitations which climate imposes on what we can profitably do for a living; rather we compound limitations by spending money to shield ourselves from the climate and thus make it necessary to produce that much more from the area where we live and work. We can irrigate or install orchard heaters, but this merely substitutes economic limits, with climate remaining at the root of both. For the most part we must accept climate as something over which we have little or no broad control and to which we must fit ourselves.

There do not seem at present to be any strong possibilities of technical developments which would in the foreseeable future enable people to control or modify climate except on the restricted scale suggested by the above examples. Experiments with artificial rain making by seeding or bombarding clouds from an airplane have so far not convinced most meteorologists that this is even potentially a technique which could alter climate (as opposed to daily weather) on a significant scale over a large area. The wind systems of the upper atmosphere, on which the broad mechanics of weather and climate appear in large part to rest, are massive forces which seem difficult if not impossible to control and at present not even easy to delineate in detail. The force or energy represented by a single hurricane or even by a normal rain storm is equal to many thou-

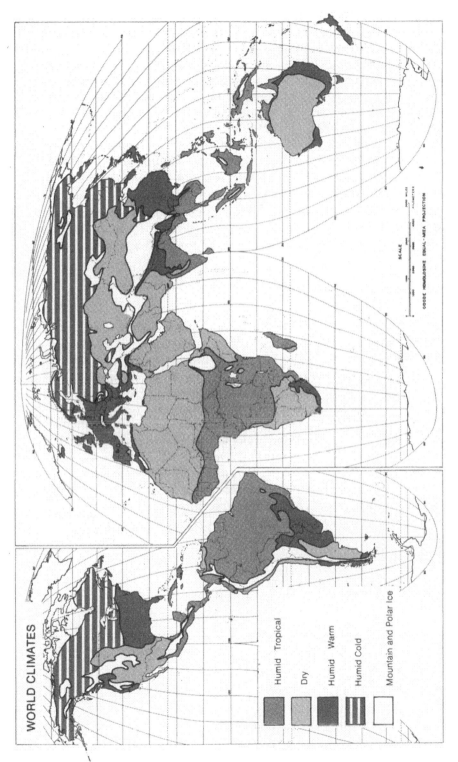

Map 11 World climatic types. *This is necessarily highly generalized so as to make it intelligible at this scale. Regional and local climates can be distinguished in a great many more categories, as in multi-color atlas climatic maps at larger scale.*

WORLD CLIMATES

Humid Tropical

Dry

Humid Warm

Humid Cold

Mountain and Polar Ice

SCALE

GOODE HOMOLOSINE EQUAL-AREA PROJECTION

sands of nuclear bombs. It is conceivable that a strategically placed and timed nuclear bomb might be used to "trigger" a desired change in weather by upsetting the local atmospheric balance on a small scale. One may expect some increasing accuracy of forecasting on the basis of statistically accumulated observations and especially with the help of weather satellites, which can improve long-range forecasting. This can be of great importance, not only to farmers but for the better periodic planning of the economy as a whole. But we do not yet begin to know enough about the atmosphere and its pattern of forces to attempt effective manipulation. Deforestation, agriculture, and atmospheric pollution alter heat radiation from the earth, increase cloud cover, and add to the particulate (dust) and carbon dioxide content of the air. These unintended side results of economic development may well change climate, and probably not for the better. They are certainly far more important and effective than any conscious human efforts to alter climate.

Climatic change and fluctuation

Climate is not constant. It moves through a hierarchy of cycles, from the obvious change between day and night (diurnal) through the changes of the seasons to longer cycles of years, centuries, and geologic periods. Between the great ice ages and the daily or seasonal changes in climate there is a gradually decreasing scale of change until one reaches a point somewhere near the beginning of the Christian era, when it can probably be said that the climate of the world as a whole became substantially the same as it is today. Fluctuations in climate on a world scale since that time have been minor compared with what went before, and such fluctuations have probably also returned periodically to a relatively constant mean. On a graph, temperature, for example, might be represented schematically as it is on Figure 12. In time the curve may drop significantly again as the earth enters a new glacial period, or it may rise significantly, perhaps as a result of man-made atmospheric pollution. But the earth does not seem to be faced with an ice age in terms of the next thousand years, and it is far enough removed in time from the last one so that the geographer

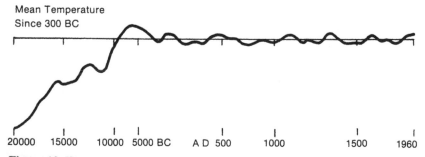

WORLD TEMPERATURE TRENDS

Figure 12 World temperature trends

needs to be concerned with it only when dealing with early people or early civilizations. This is so even though the earth may now be in what is called an interglacial period or on the threshold of a major rise in temperature or change in rainfall which would destroy or distort the average of the last two thousand years. Larger questions of climatic change in terms of thousands of years are extremely interesting, but the observable time span of what may be called human civilization is smaller. Climatic changes in the course of the last glaciation and its retreat were of transcendent importance for humans as they then existed and as they developed. It is a suggestive idea that *Homo sapiens* owes some of the flexibility and adaptability which have enabled them to multiply and spread over the world and to use the physical environment so effectively to the fact that early in its development the human species was obliged to adjust to drastic changes in climate, and that it did so successfully where many other species failed. But for most of recorded human history it can be assumed that despite continual fluctuations, the climate of the world as a whole has been broadly consistent.

Climatic fluctuations have obviously been important to people, even though in the long run fluctuations have tended to balance each other out during the last two or three thousand years. Especially in areas whose normal climate is marginal – semi-arid or semi-boreal (cold) – a slight change in rainfall or temperature can be crucial, enough to damage agriculture, for example, or temporarily to remove some of the basis for human settlement. Such fluctuations have attracted the attention of many people who have used them as evidence of climatic change, which they then apply to explain changes in human society. But it is the nature of climate to fluctuate. Progressive and consistent change is difficult to judge; it depends on the time interval used. The change from winter to summer is progressive in a sense, and cold years may be followed by warm years. But over a century or two climate is more likely to fluctuate around a relatively constant average. During the past two thousand years, alternating series of wet and dry or cold and warm years, decades, and centuries have probably not represented enough progressive change, on the basis of the scattered and unsatisfactory evidence which is available, to alter fundamentally the climate of most areas, or to provide a satisfactory basis for explaining most of the changes in human settlement during that period. There are very few weather or climatic records available before about AD 1750; earlier records which have been preserved are not easy to interpret accurately because the standards of observation or measurement are seldom clear or consistent. Geologic evidence is largely useless on a time scale of centuries, and even the evidence of changes in fauna and flora is not easy to read with precision. Other factors than climate have usually been far more important in explaining human changes.

Climate and human development

Climatic determinism has also been used repeatedly to explain current human differences over the world. Climate has been more abused than most of the environmental elements in this respect, because its force is often so important and often so clear that it tempts the investigator into simple, single-factor explanations. The Greeks, including Aristotle, Herodotus, and others, wrote of the determining effects of climate on the nature of political organization, individual temperament, morals, and tastes. Aristotle stated that the same latitudes anywhere in the world produced the same things, and that the lowest latitudes were the most richly productive, especially in precious metals. This was one of the reasons why Columbus followed a southerly route from Spain to the New World. Within the Mediterranean Basin or near it the Greeks were familiar with a variety of climates. Climate was the most obvious and to them the most important difference between the various areas which they knew, and they ascribed to climate the equally great human differences which they observed. Climatic determinism has been a persistent idea through the Middle Ages in Europe and into the modern period, when people still speak poetically of "climes" when they really mean places or areas or cultures. Several modern writers have attempted to explain differential human development and even areal differences in human temperament and abilities in terms of climate. It is suggested, for instance, that the greatest economic and intellectual development and the home of "progress" is in the "temperate" lands like northwest Europe and northern United States. It is also explicitly stated that in less "temperate" climates there has not only been less development, but human energies and mental powers, dictated by climate, do not allow it.

This would be a laughable thesis if it had not been taken so seriously by so many people. It has the appeal not only of a simple solution to a complex problem, but of a not very subtle flattery of those who happen to live in what they call the "temperate" lands and an implied insult to everybody else. It is certainly a blatant example of egoism or ethnocentrism; at best it applies to all areas the provincial standards of one. It may be true that the European, for example, is sluggish in India or in China, but this does not mean that the Indian or the Chinese is sluggish, and no objective measures could deny that both of these areas produced civilizations which were more advanced than Europe's outside the Mediterranean until about two hundred years ago. From the Asian point of view, or seen from the genuinely temperate tropics, Europe and North America are neither temperate climatically, nor, until the last two or three centuries, particularly progressive. Few people have suggested that climatic change is responsible for the recent relative decline of the East, but there have been suggestions that north Europe's relative backwardness until the modern period was the result of a colder and wetter climate which has since improved.

Evidence for this, in terms of progressive and consistent change, is slim. But more important, there have been other factors at work which adequately account for the change in Europe: notably the industrial revolution since the eighteenth century, whose occurrence was recent and was certainly not due to climatic change, and the earlier Age of the Discoveries, which brought about the readjustments discussed in Chapter 3. In contemporary terms, Europe's "progress" is said to result from the stimulating effects of its climate on human inventiveness and drive. But this is clearly reasoning after the event of the discovery and use of industrial resources in Europe and of the advantages of location which became operative after the Renaissance.

If climate were such a basic determinant, how is one to account for the fact that California was inhabited by one of the most primitive of all the American Indian groups? That northeast United States, with a climate similar to north Europe's was inhabited by stone-age people until the seventeenth century AD? That just as brilliant modern technicians, scientists, poets, scholars, and statesmen can be found in tropical countries as in the "northern temperate lands"? Is it true that different people and racial groups have different tolerances for heat and cold, different dietary requirements, metabolisms, blood pressures, and physical sizes, and also that anyone is less vigorous mentally or physically if he is infested with intestinal parasites or malaria or lives on an inadequate diet. While these things are clearly related to climate, diseases or dietary inadequacies can be removed or greatly reduced, as they have been it the last two hundred years in the West. There are no substantial physiological grounds for assuming that an Indian, for example, with proper diet and health, is not or cannot be as physically and mentally vigorous in his climate as the European or American in his.

There are good reasons for the present gap in economic development between northwest Europe or North America and the tropics of Asia or South America. Climate has something to do with this gap, not because it directly limits people's minds or physical capacities but because under certain circumstances it limits economic opportunity. These limiting effects can be and are being lessened by technical development. While it seems unlikely that India, for example, will in the foreseeable future equal or surpass the income levels or per capita economic productivity of North America or northwest Europe, the reasons for this are only partly climatic. More importantly, they are a reflection of other resources, including location, and of a huge and still growing population.

Climate cannot be used to explain everything about regional differences, nor can it safely be regarded as a direct or consistent influence on human capacities by any universal measure of accomplishment. Such a mechanistic type of analysis is intrinsically to be distrusted and is easily disproved. But this is not to deny the importance of climate. The climatic influence is less direct or perhaps less personal. But indirectly, in terms of the economic opportunity or special advantages which climate may offer or deny, it is a basic factor.

NOTES

1. See Chapter 15 for a discussion and definition of "industrial revolution."

8 THE WORKINGS OF WEATHER AND CLIMATE

There are two principal components of climate[1]: energy (usually measured in terms of temperature) and moisture (including water vapor, solid and liquid water, and the processes of condensation, precipitation, and evaporation). Precipitation may occur as mist, sleet, snow, or hail, as well as rain, but evaporation is equally important. Annual averages are the most commonly used measures for each of these components. But the seasonal distribution of temperature and precipitation, the nature of the precipitation, and the variability in temperature and precipitation from year to year may be even more important. Obviously, an area which gets most of its rainfall in winter, or in only one season, is less well off than one whose rainfall is more evenly distributed throughout the year or which has most rainfall during the growing season. The nature of the precipitation is also important. A slow gentle shower, most of which can be absorbed by the soil, is much more desirable than a violent downpour, which erodes the soil, may damage crops, and may run off the surface into stream channels before it has a chance to soak into the ground. An area with twenty inches of gentle rain a year may be much better watered than one with forty inches of violent showers.

Annual variability, especially in precipitation, is usually highest in areas where its total average amount is least. In the marginal areas where rainfall is most crucial, average annual variability may be as high as 30 to 40 per cent, as it is for instance in much of north China. With an average annual rainfall of twenty inches, thirty or more may fall in one year (which may wash away the crops or cause floods), and ten or less in the next year. Where the annual averages are higher, proportional variability is usually much less but may still be a major problem. Plants, animals, and people adjust to long-term averages of climate. They may be seriously upset by shorter-term fluctuations, or where such fluctuations are great and repeated, plant, animal, and human life must adjust accordingly and be limited accordingly. Variation in seasonal incidence can also be critical. An unseasonably cold or warm spring has widespread and usually harmful effects on all forms of life. Summer rains which on an average come early in the growing season may arrive too late or may be harmful or useless because they come too early. The extremes of climate may thus be more important than the averages.

Temperature

Temperature is important mainly for its extremes. The length of the frost-free period, between the last killing frost in spring and the first in autumn (which of course varies from year to year, with the greatest variability toward the marginal or more limited areas, as with precipitation), is probably the most important measure of the effects of temperature. The *growing season* is a little shorter, since for most plants the minimum vegetative temperature is several degrees above 32° Fahrenheit (0° Centigrade). Temperatures during the growing season are also important, and the amount of energy thus available is directly related to plant growth. Areas free from frost at all seasons have a great vegetational, agricultural, and general economic advantage; the longer the frost-free period, the less the degree of frost, and the greater amount of energy in heat terms, and of sunshine, other things being equal, the greater the advantage.[2] Extremes of heat and freezing cold of course affect many things outside agriculture, including transportation, housing, and manufacturing costs.

Temperatures also have a direct bearing on the effectiveness of precipitation, because they help to determine evaporation rates and the degree of atmospheric moisture, or humidity. Low humidity may have serious desiccating effects, and high humidity may prevent desiccation, quite apart from actual precipitation. In a hot place like India, for example, twenty inches of rainfall per year may still leave an area semi-arid, whereas parts of England which are obviously well watered receive about the same amount. The difference in effectiveness in this case is also accounted for in part by the fact that India's climate is monsoonal and that the rainfall is therefore heavily concentrated in one season (summer). Generally, the lower the temperatures (aside from seasonal variability), the less rainfall is needed to produce a given amount of moisture in the soil. Mark Twain once defined a cool day in India as one on which "the door knobs do not actually melt, but merely become mushy;" he might have added that on such days the rain does not actually turn to steam as it falls but waits until it strikes the ground.

Precipitation

Rainfall, or precipitation in general, is the more important of these two

Figure 13 Cyclonic precipitation. *This diagram was prepared from an actual weather map for a day in April in the north central part of the United States Midwest. High pressure areas of relatively colder air are indicated by H; low pressure areas of relatively warmer air by L. Winds, or exchanges of air, blow from high to low pressure areas. Their velocity depends on the* pressure gradient, *which is the result of the difference in pressure between the high and the low plus the distance between the two air masses. The fronts indicated on the weather map and in the three cross-section diagrams are discontinuity zones where cold and warm air masses meet. Sharp changes of temperature, humidity, and pressure occur along these fronts, and are often accompanied by rapid shifts in wind direction and velocity.*

components of climate, though it might be hard to choose between the Sahara Desert and the poles as rival samples of what each can produce at the extremes. Precipitation can occur only through the cooling of a moist air mass and in no other way. As an air mass is cooled, it loses some of its capacity to retain moisture, which is precipitated out of it by condensation. The amount of water vapor which an air mass can hold depends on the temperature of the air mass and increases as the temperature rises. There are three principal ways in which an air mass is cooled, and one type of rainfall is associated with each (see Figures 13, 14 and 15). Any one storm may involve all three of these processes, although they are often distinct.

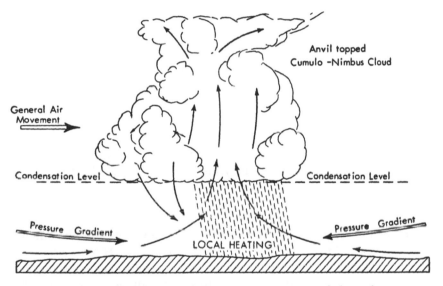

Figure 14 Convectional precipitation. *Air temperature at and above the condensation level determines whether precipitation is in the form of water (rain) or snow.* Pressure gradient *is defined under Figure 13.*

(1) *Cyclonic or Frontal.* The gradual convergence and contact of a warm air mass (rising air, low *barometric* pressure) and a cold air mass (descending air, high *barometric* pressure) along an extended front may result in the cooling of the warm air mass in one of two ways. (*a*) The cold mass, being denser, slides under the warm mass, which is thereby forced to rise over it, usually producing heavy showers. (*b*) The warm mass overtakes a cold mass and slowly rises over it; rainfall produced in this way usually falls as slow, gentle showers, often extended through one to three days. Cyclonic or frontal rainfall is characteristic of the middle latitudes, including most of North America and Europe, where warm and cold air masses from south and north respectively converge in low-pressure cells which move slowly across the land from west to east. Frontal showers may, however, occur in other latitudes, wherever warm and cold air

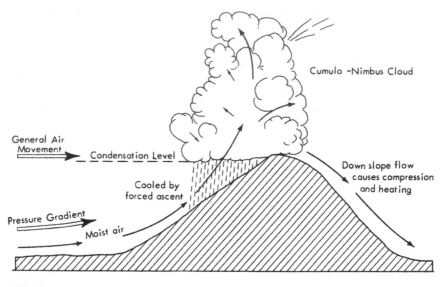

Figure 15 Orographic precipitation. *Air temperature at and above the condensation level determines whether precipitation is in the form of water (rain) or snow.*

masses meet, and in any latitude the convergence of air masses whose temperatures differ markedly may also be accompanied by relatively violent showers and by turbulence and thunder.

(2) *Convectional.* The rapid rising of air which has been heated by direct or indirect contact with the surface of the ground may cause the equally rapid precipitation of the moisture which it contains. Hot air rises, and when an already warm body of air which may therefore hold much moisture is further heated quickly, it soon rises high enough to condense most of its moisture at once, which falls as a violent shower. (The cooling of air as it rises is explained below.) This is the basis of thunderstorms, which are generally limited to places or to seasons which are hot. In large areas of the tropics they are the predominant source of rainfall, and in colder latitudes they are concentrated in summer or are absent.

(3) *Orographic.* The rising of an air mass as it strikes a topographic barrier may result in precipitation. Air cools as it rises because of the decrease in *atmospheric* pressure as altitude increases and because the surrounding upper air is usually cooler. (The greater the pressure on any gas, the higher its temperature, other things being equal; cooling produces precipitation.) There is usually heavy rainfall on the windward slopes of mountain ranges if they lie in the path of moisture-bearing winds and a much drier area on the leeward side, often a desert. In the United States the best example of this phenomenon is provided by the coastal ranges of Washington, Oregon, and California; on their seaward slopes annual rainfall is generally above sixty inches and in several places

exceeds a hundred inches; the landward side lies in what is called a *rain shadow*, and annual rainfall is under twenty inches. In large areas it is less than ten, a semi-arid or desert country. The mountains have drawn a sharp climatic line which is readily apparent in traveling across the Cascades or the Sierras. Air masses, which in this latitude are moving mainly from west to east, are milked of their moisture by the forced rise over the mountains. As they descend on the leeward side, they are warmed in the descent by rising atmospheric pressure and may often absorb more moisture than they release. This is the origin of the *foehn* or *chinook* winds, which often melt snow and cause avalanches or may parch a landscape. Similar situations, including rain shadows, are to be found wherever mountains lie across the path of the prevailing moist winds.

Climatic types

Climates may schematically be divided according to latitude, largely in accordance with the amount (hours) and directness (angle) of the sunshine which each area receives. Sunshine, or insolation, varies in amount and in intensity according to latitude because of the inclination of the earth on its axis with respect to the plane of the ecliptic as shown in Figure 17. Climates may also be classified as maritime or continental on the basis of their relationship to the sea. Briefly stated, similar climates tend to occur in similar latitudinal and continental positions. For example, northwest United States and northwest Europe have mild moist climates which are roughly the same. Both are at similar latitudes and both are strongly influenced by the sea; both are examples of maritime climates. Central Asia and Kansas are dry or semi-arid and have cold winters and hot summers; both are at similar latitudes far from the sea, and both are examples of continental climates. Latitude and continental position are far from being the only factors affecting climate, although they are probably the most important. Any division or classification based only on these two measures is grossly oversimplified. Maps of world climates use a classification system based on many different factors, which are clearly indicated. With such a system and map to refer to periodically, this discussion can deal in broader terms which, to begin with, may be clearer. The effect of latitude on climate is clear enough, and most readers will have had personal experience with it, within North America or elsewhere. The effects of the sea, which are responsible for the difference between maritime and continental climates and for many other climatic aspects, require some discussion.

Climatic effects of the sea

The sea is a great modifier of temperatures, as well as being the source of most rainfall and atmospheric moisture. Water heats and cools much more slowly than land (partly because it circulates), whereas it is only the surface or surface layers on the land which are involved in seasonal temperature changes. Thus in winter the sea still retains some of the previous summer's warmth and in summer loses the winter's chill very slowly. It doesn't fluctuate in its temperature throughout the year nearly as much as the land. Anyone who lives by the sea or by a large lake has plenty of evidence of this and of the climatic effects of the sea on the land, making it cooler in summer and warmer in winter than areas farther inland. But beyond the littoral zone (the area immediately bordering the sea), these moderating influences may be carried landward by winds; where predominantly sea-to-land winds blow without meeting major topographic obstacles, they tend to create a mild moist climate which is called *maritime* or *marine*, characterized mainly by gentle rainfall fairly evenly spaced through the year and by relatively mild winters and cool summers. This is the climate of northwest Europe, where the great European plain allows the influence of the sea to extend several hundred miles inland. In northwest United States and Canada, however, mountain ranges lie not far back from the coast, and the maritime climate is therefore limited to a rather narrow zone, as it is for the same reasons in southwest South America. Notice on the climatic map that this maritime type of climate is also present in part of South Africa, in southeast Australia, and in New Zealand. This suggests that marine climate is associated with latitude, as well as with the sea and prevailing winds.

Wind systems and climate

For reasons which need not be further discussed here,[3] there is a more or less orderly global system of winds, with roughly consistent directions in each zone of latitude (see Figure 16). Between about latitude 30° and latitude 60°, both north and south of the equator, the winds tend to blow predominantly from west to east and are called *prevailing westerlies*. Once they reach the land, especially if the land surface is broken up by mountains, their surface direction is much less consistent and may often be reversed, but at altitudes of five thousand feet or more they retain their predominantly easterly flow. If an area lies on the west side of a continent in these latitudes, its climate will therefore be influenced largely by winds of maritime origin. Notice for instance that at latitude 55° north lie coastal British Columbia, Ireland, and southern Denmark, all areas of warm winters and cool summers, while at the same latitude in east-coast North America lies northern Labrador and in east Asia, equally frigid Siberia. The maritime climates of southeast Africa and southeast Australia are, however, on east coasts; here the winds are blowing predominantly from east to west. These areas lie far enough toward the equator to be in the zone of the *trade winds*, and east to west is their principal direction. A little farther north on these east coasts, however, where the trade winds are still blowing off the sea, the climate is too much influenced by a strong, direct sun to be considered moderate.

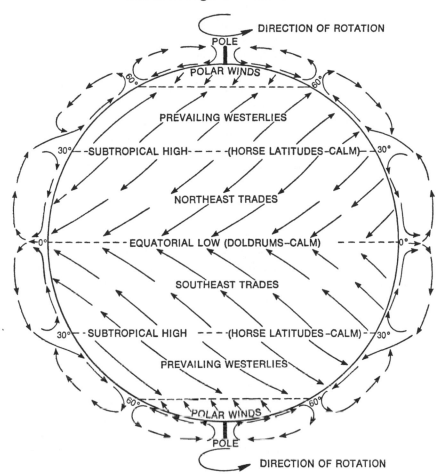

Major wind belts as they would appear on a rotating earth with a uniform surface. Winds are deflected to their right in the northern hemisphere and to their left in the southern hemisphere. Pressure and wind belts move north and south with the sun, according to season.

Figure 16 Generalized wind belts of the world

With these two exceptions, east coasts outside the equatorial and tropical zones, where the sun dominates, and outside the arctic, where the sun is weak, lie in the path of the prevailing westerlies. Their climate in a sense is imported to them from the great land mass which lies at their backs. The sea cannot affect them nearly as strongly, since the winds are blowing from the land to the sea, and these east coasts consequently have much colder winters and hotter summers than corresponding latitudes on west coasts. The interior of the great continental land masses is an area of temperature extremes, far removed from the sea, where winters are long and cold and summers scorchingly hot. Kansas, the Dakotas, or Nebraska in the United States, or Soviet central Asia, are typical samples of *continental* climates, with a tremendous temperature range between winter and summer and only a brief part of the year which could be considered moderate between the changing of the seasons. The lowest temperatures, as a matter of fact, were until recently thought to be not as one might expect at the north or south poles, but in northeastern Siberia, because it is much farther from the sea, frozen or open, than either of the poles. Summer temperatures in this area may also be uncomfortably high (unlike the poles), with a July average in the seventies and extremes in the nineties. In the zone of the prevailing westerlies, these continental interiors are the predominant influence on the east coasts, which helps to explain the rigorous winters of New England even though it is far south of London, which has mild winters.

The monsoon

The fact that the continental interiors have such marked seasonal temperature differences also contributes to the *monsoon*. In summer heated air rises over the center of the great land masses, and cooler air is drawn in from the surrounding sea to take its place. If this marine air is cooled after it strikes the land, especially if it is forced to rise by mountains or hills, or if it meets a cold air mass, heavy rainfall results. In winter the main direction of flow is reversed. The cold, dense air of the interior sinks, and winds blow out toward the sea, bringing little or no moisture with them since they originate in a very dry place and are cold to start with. The rainfall associated with the inblowing summer monsoon rarely penetrates into the interior; the winds are milked dry by mountains or lose their moisture or their force in other ways before they reach the great interior deserts. Monsoonal winds are most clearly developed in the largest land mass, Eurasia, as one would expect, where the pull of rising heated air in summer and the push of falling cold air in winter are particularly strong. Only in the Asian part of this land mass is there a large enough area far enough from the sea to establish this dominant monsoonal pattern. In Europe the sea is too close, but in parts of east Africa, where the continent bulges widest, some monsoonal effect develops, although it is usually dominated by the more powerful and adjacent Asian monsoon.

In the other continents monsoonal patterns are not readily apparent, principally because they are not big enough land masses, but the differential heating and cooling between land and sea does allow marine influences to penetrate farther inland in summer than in winter, especially on east coasts. This is part of the reason for the hot, sticky summer days in

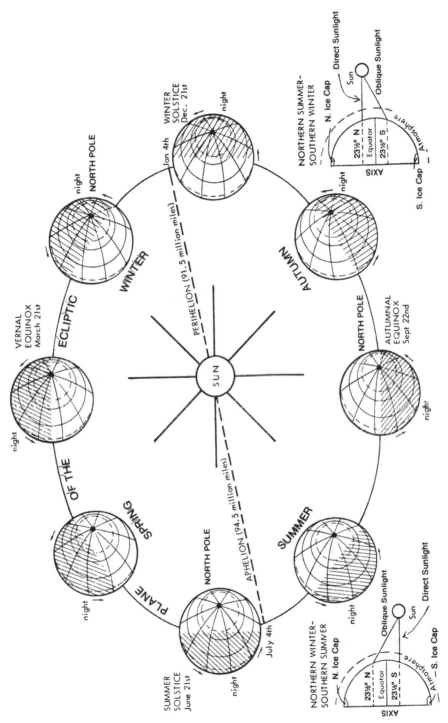

Figure 17 Earth-sun relationships

places like Washington, D.C., or New York; some of their uncomfortable summer weather is imported from the tropical Caribbean. Even in Asia, however, the actual forces involved in the weather systems which are broadly labelled monsoonal are very much more complex than this simple account suggests and are by no means fully understood by meteorologists. It has been said that "every schoolboy understands the monsoon, but the Indian Meteorological Service is still puzzled by it."

The seasons

Seasonal changes result from the 23.5° inclination of the earth on its axis in relation to the orbit of its course around the sun (the plane of the ecliptic), as illustrated in Figure 17. The sun's position in the sky in relation to the earth, and hence the angle at which its rays strike the earth, migrates annually north and south of the equator. The sun appears directly overhead at noon, and its rays therefore reach the surface of the earth directly, only between the Tropics of Cancer in the northern hemisphere and Capricorn in the southern hemisphere. June 21 marks the farthest point of the sun's apparent northerly migration, when its noon rays are directly overhead at the Tropic of Cancer, 23.5° north of the equator. On December 21 its southerly migration reaches the Tropic of Capricorn at 23.5° south. The zone between these two lines, or the *tropics*, thus receives maximum insolation which varies little throughout the year and is continuously warm with little or no seasonal changes in temperature. Beyond the tropics insolation and the relative lengths of day and night not only vary from season to season but heating effectiveness is greatly affected by the angle at which the sun's rays penetrate the atmosphere. The atmosphere acts as a powerful insulating blanket. The difference between the result produced by vertical insolation and by insolation at an angle is great enough to account for a large part of seasonal changes. This is readily demonstrated by observing the difference in heating effectiveness between the sun at noon and at dawn or dusk.

Climate and economic opportunity

This highly generalized summary of the mechanics of climate should not discourage examination of climatic processes and regional differences in more detail. A good starting point is provided in any map of world climates. But this discussion is concerned more importantly with the effects of climate on human society. Having discarded the thesis of climatic determinism, and having agreed also that at the extremes, such as the desert or the arctic, climate's effects are too obvious to require discussion, what scope is left for the climatic factor? As already suggested, it may be clearest and most accurate to think of climate in terms of the kind and degree of economic opportunity which it offers.

This concept most nearly fits the correlation between climatic areas and areas of human settlement or of economic activity and does not inveigle us into considering the effect of climate on people's minds.

In monsoon Asia[4]

The climate of monsoon Asia, for example, offers an opportunity for intensive agriculture based on wet rice. In this area of long growing season, high summer temperatures, and generally adequate but not excessive rainfall, rice can yield more food (or calories) per acre than is produced by any other cereal crop or by any other area of comparable size. The connection between this climatic advantage and the very large and dense populations of monsoon Asia is clear, in terms of economic opportunity. Irrigation makes almost any agriculture more productive. Monsoon Asia's rainfall is seasonally concentrated in summer and is highly variable from year to year in amount and in seasonal incidence, so that irrigation is especially necessary. Rice is the dominant crop, which means high water requirements. Irrigation became a fundamental basis of most agriculture in monsoon Asia, and the complex social, administrative, and economic organization necessary to maintain irrigation systems has contributed to the general nature of most Asian civilizations, especially to the strong role of the state, or at least of organized groups, and the relatively weak position of the individual entrepreneur. The productive nature and special form of their agricultural systems lie at the root of all the great Asian civilizations and have done much to influence their character and to distinguish them from other civilizations elsewhere. Climate has not made the Asian mind work differently from the European, but it has provided a different kind of economic opportunity. This in itself tended to encourage the growth of a distinctly Asian culture. Climate was not of course the only factor, but in this case its role, in a permissive sense, is apparent.

In the steppe

The semi-arid areas of the world are in a different climatic situation. *Steppe* is used loosely as both a climatic and a vegetation term. Climatically it applies to the semi-arid areas, transitional between desert and humid, too dry for any significant tree cover but moist enough to support grasses or scrub. Steppe covers the Great Plains of the United States, most of the Prairie Provinces of Canada, part of the Argentine *pampa*, a narrow rim around the great Australian and African deserts, and most of middle and southern central Asia. Here climate offers a relatively limited economic opportunity, and of a particular kind. Except in the relatively small areas where irrigation is feasible, or where there are particular local opportunities such as minerals, the steppe does not support a large or dense population. Most of the people who live there earn their living *extensively*, by applying relatively small amounts of labor or capital to proportionately large amounts of land. The climate, except in special situations, does not make a heavy investment of labor or capital profitable because the return would not be great enough to compensate for the expense, as it is for instance in monsoon Asia or New Jersey, where land is worked *intensively*. Steppe areas therefore tend to be characterized by sparse populations engaged in growing drought-tolerant crops like wheat on very large farms in the best or least dry sections, or in the raising of animals – cattle, and in the drier or hillier parts, sheep or goats.

If mineral resources are fortuitously present, mining may supplement the economy and attendant cities may grow up as trade, transport and service centers. Denver is a good example. Denver also profits from location where plains give way to mountains and where transport routes bunch together to get across by the easiest route. Bulk is commonly broken there, and Denver makes a good assembly and distribution center, as well as a good operations base for railroads and airlines, all of which provide basic employment for Denver's population and in turn support local service industries and service employment. Denver would be a much smaller place, however, if it were not that it lies between two much more densely settled and productive but complementary areas, eastern United States and the west coast. Exchange between them takes place across the intervening steppe, through Denver. The large amount of intervening space, plus the barrier of the mountains, makes a trade service center and break-in-bulk point useful along the east-west route. The same set of factors is involved in the pattern of trade service centers across the steppe of Eurasia and on its margins. But manufacturing in the steppe is unlikely, aside from the primary processing of local raw materials such as ores, the supply of the small local markets, or a few specialized industries. The major markets will be elsewhere, where the people live, in better watered places.[5] The steppe is also likely to experience considerable climatic fluctuation, especially in rainfall, since high variability is unfortunately associated with low annual averages. In a commercialized economy like the United States, fluctuating rainfall may be accompanied by waves of prosperity and depression in the steppe, with attendant political unrest or support for radical policies of relief or reform. In a pre-commercial or nomadic economy such as central Asia's has been until recently, climatic fluctuations may contribute to conflict among nomadic groups, or between the nomads and the sedentary farming cultures around them.

Climate and rail patterns

In the United States the pattern of transport routes reflects this climatically induced economic nature of the steppe. A map of railroads (Map 12) makes this point unmistakable. At about the hundredth meridian of longitude, the rail pattern in the United States contracts more or less abruptly and sends west into and across the plains only a few lines, in sharp contrast to the dense network to the east. At first sight this might

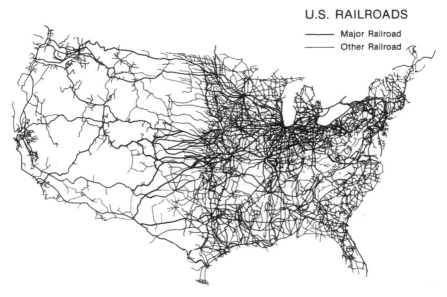

U.S. RAILROADS

———— Major Railroad
———— Other Railroad

Map 12 United States railroads. *Compare with the rainfall map in Figure 5 on page 35.*

appear to be due to the mountains, especially if one looks at a generalized physical map of the United States. Actually, the area concerned is the Great Plains, and while there is a rise in elevation westward from the Mississippi, it is so gradual that the landscape looks virtually flat or gently rolling, and there is no real barrier of hills or mountains until the sharp front of the Rockies is reached. This is one of the disadvantages of the altitude tint (color) system of showing elevations, which is discussed in Chapter 12. On a rainfall map (Figure 5, page 35), however, a major break is shown which coincides quite closely with the hundredth meridian. This is the line (actually a zone) between steppe (which merges with a more humid but still treeless area called the prairie) and the well-watered east, between two areas of differential economic opportunity. The line of twenty inches (5080 millimeters) average annual rainfall, or the twenty-inch *isohyet*, is one of the most important regional lines in the United States. Population reflects it (Map 13), with relatively dense settlement to the east and sparse settlement to the west.

These are matters of great importance for a railroad. Transport routes in general, as pointed out in the analysis of Switzerland, react far more sensitively to the volume and direction of trade than they do to physical barriers. A railroad represents a heavy capital investment, and such

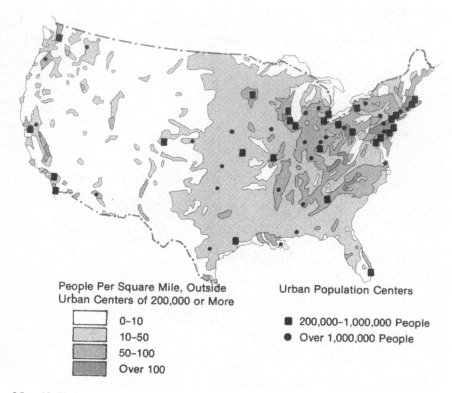

People Per Square Mile, Outside
Urban Centers of 200,000 or More

- 0–10
- 10–50
- 50–100
- Over 100

Urban Population Centers

- ■ 200,000–1,000,000 People
- ● Over 1,000,000 People

Map 13 United States population density. *Compare with the rainfall map in Figure 5 on page 35. Population data are from the 1980 Census.*

investment does not pay unless there is a large and constant volume of business. There are no important physical obstacles for a railroad in the Great Plains, but there is a clear economic problem – a small volume of commodity flow, since there are relatively few people and few large cities. Climate is largely responsible for this. The railroads in effect contract their routes and get over the empty space as economically as possible by what are called "bridge lines." The two ends of the bridge are the humid east and the humid or irrigated Pacific coast, where there are enough people and enough exchange between them to carry the railroads across the intervening relatively empty space.

On a map of North American railroads, Canada seems to deny what has been said about the hundredth meridian. To the west of it in Canada the railroads actually spread out, whereas to the east they contract into only two lines, bridging what is obviously an economically empty space before they fan out again into the humid areas of eastern Canada. Climate is responsible for only part of this, but since it has come to our attention at this point, it makes a useful demonstration not only of the climatic factor but also of the fact that the environment tends to operate as a whole as an influence on mankind. Look first at the empty space on the map. That part of Canada is an area of nearly bare crystalline rock known as the Candian Shield; most of the original soil was scraped off by the last glaciation, and what little remains is too thin and poor to support anything but a spruce/fir forest except for a few limited areas. It is an area largely avoided by human settlement, as a map of population shows, but soils are more important than climate in accounting for this. Climate in the Canadian Shield is somewhat less good for agriculture than in the Prairie Provinces farther west, despite the fact that the average annual rainfall is higher in the Canadian Shield; the rainfall is concentrated in late summer, which means that it is less useful, and the summers as a whole are shorter and cooler than in the Prairie Provinces. This is a minor point, however, compared with the virtual absence of usable soils. Actually there are large and valuable deposits of mineral ores in the Canadian Shield: nickel, iron, copper, zinc, uranium, and others. They do not provide a basis for large-scale settlement because, in the absence of other economic opportunity in this generally marginal area, the ores are shipped out to where the people live and where the major markets and manufacturing centers are. The railroads thus have to cross this empty space of the Canadian Shield as economically as possible, picking up some ores on the way, but not really running on a profitable basis until they reach southeastern Canada, where soils, climate, location, and other factors combine to provide economic opportunity for denser human settlement.

West of the shield lie the Prairie Provinces – Alberta, Saskatchewan, and Manitoba. Here the map shows markedly greater popular density than in the shield (or than at corresponding longitudes in the United States to the south), and a spreading out of the rail routes to service it. Climate has played here both a direct and an indirect role. Directly it is relevant because, while average annual rainfall is much the same for the Prairie Provinces as for the Great Plains area of the United States (which has a much sparser population), it is more effective because evaporation rates are lower in this cooler climate. It is thus possible to grow wheat, for example, on a larger basis in Alberta than in Wyoming or Colorado. A map showing generalized types of farming makes this clear and shows a skewed pattern of wheat farming extending much farther into the dry west on the north than in the south, leaving most of the Great Plains in the United States dependent on grazing, and hence supporting relatively fewer people.

Indirectly, climate has helped to account for this difference in terms of alternate opportunities. Canada has few favorable alternatives to the Prairie Provinces in terms of general economic opportunity, whereas the United States has the Middle West as an obviously superior domestic rival agriculturally to the Great Plains, and a wealth of other areas capable of supporting dense settlement on a high economic level. Most of Canada is too cold (or in the case of the shield lacks also the basic factor of workable soils) to permit large-scale settlement, and the climate of the Prairie Provinces makes it one of Canada's favored areas where it might be a poor one in the United States and would be used or settled much less fully.[6]

Climate, as repeatedly pointed out, plays a predominantly passive or permissive role. It allows certain developments rather than causing them. Accessibility, cheap exchange, local mineral deposits, economic and political organization, the availability of capital surpluses, and the growth of technology are more important than climate as advantages for trade and manufacturing. Northwest Europe remained a backward area until these other advantageous factors were developed or became effective, although the climate did not change. The relevance of the favorable climate is that without it, the other advantageous factors would be less effective or even inoperable. There is at least one respect, however, in which northwest Europe's climate has been a more direct advantage for trade and manufacturing. Cheap exchange, or cheap assembly and distribution, is a crucial factor for both of these functions. The mild, moist, maritime climate of Europe ensures that the great European plain will be traversed by many rivers, and that their navigability will be enhanced. In maritime Europe, west of Poland and south of Norway, rivers do not often freeze solid in winter, and serious floods or prolonged low-water periods are relatively infrequent, since the rivers receive rainfall quite evenly distributed throughout the year. The Rhine, for instance, is a relatively short river, but it is navigable nearly up to the mountain front of the Alps, including a short navigable stretch within Switzerland itself. This has been an important asset for a Europe so heavily dependent on cheap exchange. Water routes, by sea or by river, join together nearly all of its parts and make possible cheap movement of bulky low-value goods, like coal or foodstuffs, which are essential in large volume to maintain a modern specialized industrial economy. Internal waterways helped to make the industrial revolution possible in Europe before the building of railways, and after industrialization water routes could move bulk goods more cheaply than railways. Rivers remain a vital economic asset for Europe, and their usability is directly related to Europe's climate.

This chapter has attempted to discuss climate as an influence on human society primarily in terms of economic opportunity. It has been necessary to generalize very broadly in order to make the essential point and to focus attention on climate to the disregard of other factors. Climate, like any other influence on people, does not operate alone, but as part of a complex in which the rest of the environment, relations in space, and other separate or non-geographic factors are also involved. Climate is one of many factors lying, one may say, at the foundation of human development and giving people the material, good or bad, with which they are obliged to work.

NOTES

1. Atmospheric pressure and winds are important components of weather and climate, but they will be treated as they arise in the course of the following discussion. Climate is simply long-term average weather.

2. Heat may of course be excessive and sunshine may be restricted by humidity or cloud cover, or by day length. Much of the tropics may thus be less productive agriculturally than areas where temperature and moisture are in better balance and summer days are longer. The effective growing season is a result of a complex of factors.
3. These reasons are mainly the exchange of heat between high and low pressure areas and the rotation of the earth.
4. South and east Asia, where long warm summers are combined with usually adequate rainfall; most of India, Bangladesh, Sri Lanka, the Indo-Chinese peninsula, Indonesia, the Philippines, China, Japan, and Korea.
5. The Soviet Union has, for strategic and political reasons, located considerable manufacturing in Russian steppe areas.
6. There is a non-climatic factor at work in this part of Canada which will continue to increase its population and prosperity still further; large deposits of petroleum and natural gas (in addition to existing coal) have been discovered and are being worked there. All three are the most important deposits in Canada as a whole, although they do not rival the large United States deposits.

9 VEGETATION AND HUMAN SETTLEMENT

Natural vegetation is largely a corollary of climate, but it can much more easily be modified. It is a useful classifier, since it reflects the climate and may also reflect soils, ground water, and relief, as well as human activities. Vegetation also reflects the nuances and variations of climate from one small area to another and from season to season or year to year, so that it may often be a more reliable guide to climate than is provided by short-term records. Comparison of the world map of vegetation (Map 14) with the climatic map (Map 11, page 91) will demonstrate how closely the two co-vary in their spatial distribution. The vegetation map shows, as it indicates, "natural" vegetation, however, and in the parts of the world where people have settled in large numbers most of the original vegetation cover has been removed. The nature of the original vegetation cover has had to be reconstructed on the basis of climatic and soil conditions, plant pollens, and by means of the few surviving specimens. In some areas where people have not settled so densely, the original vegetation largely remains and is in part responsible for restricting human settlement. The most obvious and important example is the equatorial rain forest of Africa, South America, and southeast Asia, where the barrier effects of vegetation help substantially to limit economic opportunity, although the rain forest everywhere is now being heavily cut into. Vegetation, however, does not act alone. The climate of the equatorial rain forest is harsh, the soils are generally poor, and the areas as a whole cannot support enough people at a high enough economic level to make the complete removal of the rain forest physically or economically possible, although as mentioned, inroads have greatly increased. The same limits, and the recent increase in exploitation, apply to most of the wide band of coniferous forest in the northern hemisphere between about latitude 50° and latitude 70°, known as the *taiga*.

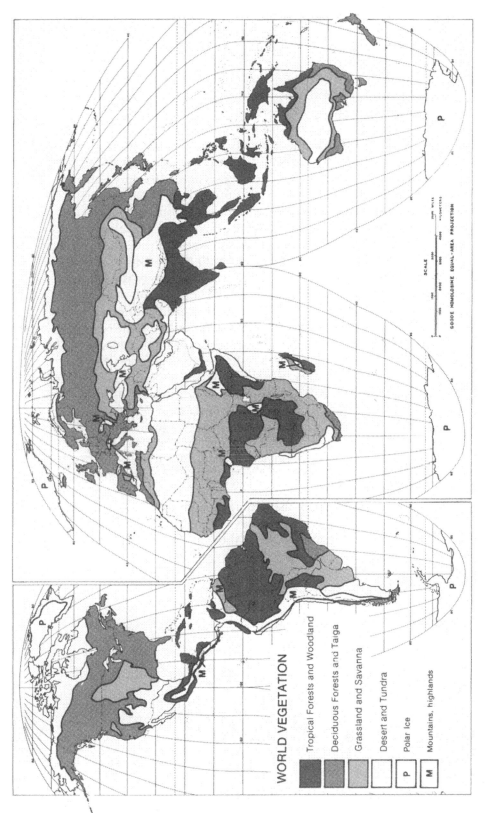

Map 14 World vegetation types. *This is necessarily highly generalized so as to make it intelligible at this scale. Regional and local vegetation complexes can be distinguished in a great many more categories, as in multi-color atlas vegetation maps at larger scale.*

WORLD VEGETATION

- Tropical Forests and Woodland
- Deciduous Forests and Taiga
- Grassland and Savanna
- Desert and Tundra
- P Polar Ice
- M Mountains, highlands

SCALE

GOODE HOMOLOSINE EQUAL-AREA PROJECTION

Barrier forests

Apart from the useful or cultivated plants, vegetation tends to act as a physical barrier in restricting human activity, although where trees are absent and grass is dominant, vegetation may offer an opportunity for movement. As already pointed out, where vegetation has been a barrier in the densely settled areas, it has largely been removed. But before this removal was accomplished, or before other conditions made the removal physically and economically possible, forests especially were an important limit on human activities. The role of the forest in European history was that of an obstacle to trade and circulation, to agriculture, and to the achievement of national unity. The removal of most of the great European forest stands at the end of the Middle Ages coincided with a great increase in trade, with the growth of towns, and with the rise of modern European civilization. The forests also acted as refuge areas for persecuted or dissident groups and as bases for outlaws or robber bands. Sometimes the forests served to shelter more organized attempts to levy tolls on traffic passing through or around them and were the seats of the so-called robber barons. Most historians believe that the story of Robin Hood is a cherished fiction, but at least there were people like him in Sherwood and in many other forests in England and Europe. Sherwood forest is almost gone now; the area where it once stood is in the modern English industrial belt. Nottingham Castle still stands, but it is surrounded by a manufacturing city, and tourists are taken to an obviously cut-over and tiny patch of forest to see the "Major Oak" under which Robin is supposed to have assembled his band. When trade and its urban bases grew large enough, and when the national states grew strong enough to put down local anarchy, most of the lowland forests of Europe were destroyed, although in the remote or hillier areas some extensive stands remain and are used as sources of timber or for recreation. An equally if not more important factor in the removal of the forests was the

Figure 18 Generalized vegetation sequence

There is a more or less wide zone of transition between each major vegetation type. Note also that the spacing of latitude lines has been foreshortened in order to expand the low and middle latitudinal areas.

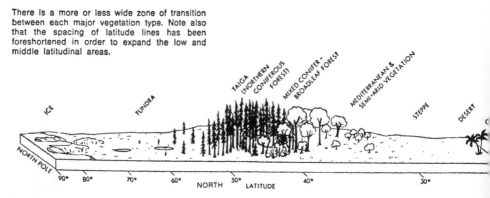

expansion of agriculture to feed the growing towns and to support an increasing rural population.

In other heavily forested parts of the world, the forest still shelters dissident groups. In Malaya, for example, it has been said that "the jungle is not neutral; it is immensely biased against the forces of law and order. The fugitive, the outlaw, the guerrilla fighter, has every advantage."[1] During the war in the South Pacific theater from 1941 to 1945 and again during the long war in Vietnam, the forest was a military factor of supreme importance. American military efforts to defoliate the forest in Vietnam with chemical sprays, to deny its cover to guerrillas, underlined its strategic importance but yielded devastating results for the human population, for whom the defoliants proved severely toxic.

There has been a continuous rivalry between people and the forest, in which the forest has often been associated psychologically with the unknown, the dark, magic, the home of feared spirits, and the symbol of mystery.

Forests in history

The earliest civilizations arose in largely treeless areas – Egypt, Mesopotamia, north China, and northwest India. Before recorded history people seem generally to have avoided dense forest, at least for their settlements. Once people were able to use and manipulate fire, perhaps as long as half a million years ago, they had a powerful weapon against the forest or against any vegetation. They also learned to use stone tools to kill trees by girdling or felling them. Human agency has thus altered vegetation on a large scale for a very long time. Burning was used to clear forests of underbrush in order to make hunting, grazing, or movement easier, or to improve pastures by favoring the types of vegetation which flourish after fires. Forests have seldom been attractive to advanced human settlement. No heavily forested area, so far as we know, has ever

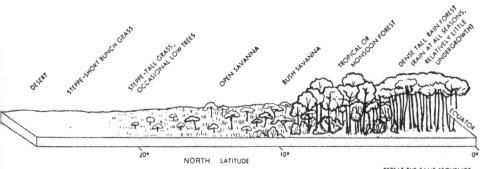

DESERT   STEPPE-SHORT BUNCH GRASS   STEPPE-TALL GRASS, OCCASIONAL LOW TREES   OPEN SAVANNA   BUSH SAVANNA   TROPICAL OR MONSOON FOREST   DENSE TALL RAIN FOREST (RAIN AT ALL SEASONS, RELATIVELY LITTLE UNDERGROWTH)   EQUATOR

20°   10°   0°
NORTH   LATITUDE

REPEAT THE SAME SEQUENCE
OF VEGETATION IN THE
SOUTHERN HEMISPHERE

been the base for a technically or economically advanced civilization, with the possible exception of the Maya culture of Central America, whose origins are obscure. It seems likely that there, as in other forested areas, sophisticated development took place only after the trees were removed.

Rivers have often been the first avenues of attack on the forests and the first corridors of denser settlement, as the only zones of easy movement within an otherwise barrier area. In Europe and Asia deforestation began along the rivers, as it did and is still doing much later in the other continents. Forests were cut into for practical uses as well as to clear the land for cultivation or trade. Wood has always been in demand as building material and fuel, and for charcoal as a more efficient fuel. Charcoal burners were often the pioneer forest destroyers, ranging widely and reducing the wood to charcoal in small kilns within the forest so as to use local fuel and to lessen the transport problem by converting a bulky, low-value product into a more compact, higher-value one. As the ravages of the charcoal burners progressively exhausted the more accessible timber stands, industries dependent on charcoal suffered, especially the smelting of ores. This problem was an important factor in the rapid use of coal on a larger scale, in the form of coke, in eleventh-century China and in seventeenth- and eighteenth-century Europe. In modern northwest Europe, deforestation took place earliest in England. In addition to the fuel problem, there was a shortage of ship timbers. Oaks large enough for the keels and stern posts of naval ships of the line were scarce by the eighteenth century. The development of metal ships came in time to solve the problem, but for a considerable period the size and number of capital ships had been to a degree limited by the decreasing availability of large oak timbers. Another important use of forests in Europe was for the grazing of pigs on beech mast (beech nuts); this represented a significant addition to food production, especially where open beech forests were accessible to densely populated areas.

In North America the forest has also played a barrier role, as a bar to the expansion of agriculture and settlement, an obstacle to movement and trade, and a base for dissident groups. Most of the taiga stands of Canada are still relatively little used, and even in the United States, although most forests have been removed, cut over, or heavily invaded, it has not been necessary or profitable to do away with all of them. But the advance across the Appalachians in this country depended on the removal of the hindrances which the forest presented.

Barrier rain forests

To find the clearest example of the barrier effects of vegetation, however, one must turn to the extreme represented by the equatorial rain forests. As pointed out above, the climate and soils of these areas are also poor, and these two factors do as much or more than the vegetation to limit human settlement. Nevertheless, the rain forest is a formidable discouragement to economic development, and the sparse populations who are able to maintain themselves within it can do so only at a very low level of development. Access, movement, and exchange are at a minimum, and regional specialization is virtually impossible. What movement does take place in the rain forest, aside from hunting and collecting, is often confined to navigable streams or rivers, the only practicable routes through such a dense wall of vegetation. Agriculture is a relatively difficult and unrewarding affair, since the massive growth of trees must first be removed, and also since the soils are generally of low fertility. Most rain forest dwellers earn their living by a mixture of shifting agriculture (see Chapter 10) and the gathering, collecting, and hunting of food in the forest: fish, roots, berries, nuts, and game. The rain forests are excellent refuge areas, and their human populations may often be composed largely of groups who have retreated or been pushed off better lands elsewhere by stronger groups. They find the protection they need in the rain forest, not only because it is hard to get at but because it is also undesirable to other people, who find greater attractions in other places. The swamps and swamp forests of Georgia and Florida in the United States are also refuge areas and support small groups of people whose culture and technological level differ sharply from American norms.

The rain forest is still partly responsible for isolation and a low level of economic development in the Congo Basin of Africa, the Amazon Basin of South America[2] and in much of southeast Asia. Most of Borneo and New Guinea, for example, are sparsely occupied by primitive subsistence groups, including, until very recent years, cannibals and head hunters, even though Borneo and New Guinea lie relatively near Java and Australia, both highly developed areas. As the vegetation map shows, the natural cover of Java was largely rain forest. But Java is small enough to be easily accessible by sea, lies in a commercially strategic location, and is fortunately supplied with fertile volcanic soil, unlike most tropical or equatorial regions. This combination of favorable factors supported a relatively dense and highly developed population which long ago removed most of the original rain forest. Along the coast of Malaya, too, where access by sea is easy, large commercial plantations have grown up at the expense of and despite the rain forest. But although local conditions, and especially access, may dent or destroy the barrier effects of the rain forest, for most of the area which it covers human activity is severely restricted.

Timber resources

Forest timber is a potentially valuable resource but not always an economically usable one. Most of the world's remaining forested areas, including the two most extensive types, the taiga and the tropical rain forests, are not used commercially on a large scale. Timber is a low-value commodity by weight; it cannot bear expensive transport hauls without becoming impossibly expensive in the market. Actually, the price for most timber in most markets represents mainly transport costs. Where water routes are available on which the logs can float or be carried to market, transport costs are lowest, and the major commercial logging areas of the world are on or near water. To haul logs very far by truck or train, let alone by porters or wagons, soon makes them prohibitively expensive unless they are of very high unit value sorts, such as teak or mahogany. This tends to mean that forested areas far from water routes are much less heavily used and may be entirely unused for commercial timber. This is especially so when such areas also lie far from land transport lines or from lines which have the high constant volume of flow associated with low transport costs. Hence much of the taiga of Canada, Scandinavia, and Russia, where population is sparse and railways few, is not commercially usable, although the soft wood of coniferous trees is valuable especially for pulp, used to make paper.

The access and transport problems in the rain forests are even worse. As one result of the constant, heavy rainfall and high year-round temperatures, a multitude of tree species thrives in the rain forest, and the trees tend to grow in a vast jumble with no one species clustered in consistent stands. By no means all trees, or all sizes of any one species, are commercially valuable, especially where transport costs are high. In the rain forest with its special problems of access and transport, only a few species are valuable enough to bear the transport costs. Since the individual mature trees of the valuable species are scattered over a wide area, logging operations are expensive. Loggers attempt to keep down costs by cutting trees on a mass wholesale basis, and they cannot cheaply wander hundreds of yards with their equipment through a rain forest from one mahogany tree to another. Tropical hardwoods like mahogany, teak, or ebony are commercially valuable because they are dense, heavy woods and have an attractive appearance when polished. They can resist rot, cracking, sea water, and fire much better than other lighter woods. Tropical hardwoods command a very high price on the market. But the weight and density which makes them valuable also make them expensive to transport. They are so dense that most of them will not float, so the rivers which are plentiful in most rain forests cannot be used to transport the timber unless the trees are first girdled and left to dry out for a year or more, or unless the logs are rafted downstream, sometimes by being lashed to lighter logs. All this takes time and costs money and helps to explain why tropical hardwoods are so expensive and so relatively little used. The great majority of the timber resources of the rain forests are commercially untouched.

In the more temperate, densely populated areas of the world where commercial logging has proceeded on a much larger basis, lumbering has been more wholesale, and huge areas have been cut off. In some cases where trees were the major local resource, the end of logging has left behind it an area of both physical and economic desolation, like a large part of the upper peninsula of Michigan. The logging industry realized long ago in Europe that selective cutting, leaving behind the smaller trees, and replanting, although more expensive in the short run, was in the long run wiser planning than a wholesale removal of trees. This is true not only for the long-term yield of timber, but for the prevention of erosion and floods. Such practices are belatedly spreading in the United States, but not before large areas of forests have already been ruined by wasteful logging which aimed only at the short-run or immediate gain. Even under the best conditions of "tree farming," however, lumbering is necessarily an *extensive* industry, applying relatively small amounts of capital or labor to relatively large amounts of land. Regions where lumbering is an important element of the economy are in part for this reason relatively sparsely populated even though they may be productive in total yield of commodities for export.

The grasslands – zones of movement

The role of vegetation in offering an opportunity for movement is exemplified by the steppe, the great grasslands. There are no trees to act as barriers to movement, although the water problem may be serious. Most steppe grasses have very deep roots and during the long dry spells their surface leaves die off and turn brown. With the reappearance of rain, they come to life again. Trees cannot do this; they may survive dormant during the winter or during a dry season, but once their leaves and branches are killed, they die. Most trees are relatively shallow-rooted compared with most steppe grasses, and they require larger amounts of rainfall or of near-surface ground water; in general, they cannot survive outside the humid parts of the world. This can clearly be seen by comparing the world maps of climate and vegetation.

Steppe rainfall is often not enough to support agriculture without irrigation, or rainfall deficiencies may make agriculture difficult or precarious. The rainfall will, however, support grass, and the grass can support animals which in turn support man. Aside from the sedentary commercial farmers (a quite recent development in semi-arid areas, dependent on cheap world-wide transportation and on other technological and economic changes), those who live in the steppe must of necessity be in continual motion, in pursuit of grass for their animals. The American cowboy and the Argentinian gaucho, in a commercial economy, do have more or less fixed residences even though they spend much of their time moving about on the range. But in the great steppe area of Eurasia, just as in the grasslands of North and South America until a hundred years ago, most of the people have until recently been nomadic. They were not importantly dependent on markets and did not have to prepare their products for shipment, since they largely consumed what they produced. Their home was a temporary one, wherever the search for forage and water for their animals led them. Movement was relatively

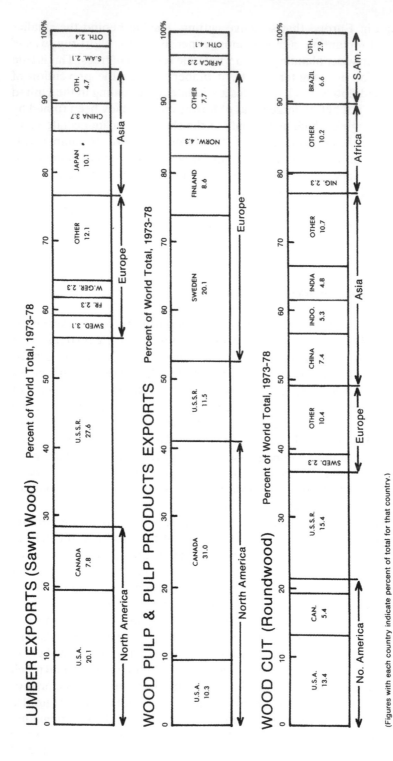

LUMBER EXPORTS (Sawn Wood)   Percent of World Total, 1973-78

| U.S.A. 20.1 | CANADA 7.8 | U.S.S.R. 27.6 | SWED. 3.1 | FR. 2.3 | W.GER. 2.3 | OTHER 12.1 | JAPAN 10.1 | CHINA 3.7 | OTH. 4.7 | S.AM. 2.1 | OTH. 2.4 |

North America — Europe — Asia

WOOD PULP & PULP PRODUCTS EXPORTS   Percent of World Total, 1973-78

| U.S.A. 10.3 | CANADA 31.0 | U.S.S.R. 11.5 | SWEDEN 20.1 | FINLAND 8.6 | NORW. 4.3 | OTHER 7.7 | AFRICA 2.3 | OTH. 4.1 |

North America — Europe

WOOD CUT (Roundwood)   Percent of World Total, 1973-78

| U.S.A. 13.4 | CAN. 5.4 | U.S.S.R. 15.4 | SWED. 2.3 | OTHER 10.4 | CHINA 7.4 | INDO. 5.3 | INDIA 4.8 | OTHER 10.7 | NIG. 2.3 | OTHER 10.2 | BRAZIL 6.6 | OTH. 2.9 |

No. America — Europe — Asia — Africa — S.Am.

(Figures with each country indicate percent of total for that country.)

Figure 19 World lumber, pulp, and wood cut

easy, since along the steppe or at intervals pasture and water are available. Movement is necessary because the grasses or water cannot maintain a herd in one place for long; the animals must move on before they exhaust the often sparse pasturage.

Nomadism is on the wane now, as the world economy and global political pressures impinge upon it. Most of the great steppe of Eurasia is politically part of the Soviet Union, and most of the rest belongs politically to China. For both countries it is highly inconvenient to have subject populations moving about with the freedom of the nomads, and the governments are also anxious to derive some economic advantage from the steppe areas which the nomads control. Over much of this area nomads are being persuaded or forced to adopt fixed residences and to produce animal products for marketing, or to take a job in a meat-processing plant. A few of them, like a group of the Kazaks, have rebelled and fled to other countries. There are still a few essentially nomadic areas left, but until a few years ago this whole vast region was a zone of continual movement. There are still a few essentially nomadic areas left, but until a few years ago this whole vast region was a zone of continual movement. Annual migrations in search of pasturage might take a given group or tribe several hundred miles.

The nomads could also be highly mobile from a military point of view. The typical nomad doubled as a mounted tender of flocks and as a mounted warrior. He usually had a telling advantage, despite his small numbers, over the sedentary empires which lay around him, because he could strike hard without warning, quickly retire into the steppe, and if necessary strike again a hundred miles away the following day. The nomad inhabited or wandered over a poor land, and he was perennially attracted by the comparative riches and easy plunder of the better watered areas around him.

Occasionally the individualistic nomad would submerge his clannish or independent tendencies in a large organization or alliance of steppe groups and establish a military empire which swept everything before it, until in a few generations the empire dissolved into the more normal anarchy of the steppe. By such conquests, and by their usual pastoral migrations, the nomads were often carriers or transmitters of culture. In Eurasia their role in this respect was particularly important, since the steppe runs virtually unbroken from the eastern margins of Europe to the confines of the great Oriental civilizations. Contact between these two cultural and economic centers of Europe and east Asia, especially before the development of sea routes, depended on the nomads. The main route of trade between east and west followed the Eurasian steppe, and ideas flowed with trade.

Human modification of vegetation

Human modification or removal of the natural vegetation may often produce more far-reaching effects than the original cover exerted. The original cover, before it is modified by people or by animals kept or encouraged by people, is the result of ages of natural selection and adaptation. This is a continuing process in which vegetation changes as climate changes or as the total pressures of animal and vegetable life change, but it can easily be upset. To remove the vegetation, or to remove one element of it, destroys what is fittingly called the balance of nature. Although this may be worth doing in terms of human benefit, it may in some cases bring results which were not anticipated and which are the reverse of beneficial. Harmful results are most pronounced in the form of erosion and floods. Vegetation, especially trees, acts like a sponge on the land. It breaks the force of rainfall, absorbs moisture, holds the soil in place, and like a reservoir, it releases moisture from its sponge of roots and soil only gradually. With the vegetation removed, a heavy shower, especially on a slope, may wash away the soil and may run off the land and into stream channels too quickly to be absorbed adequately. This contributes to worse and more frequent floods and is one of the factors responsible for troubles with the Missouri and Mississippi rivers in the United States, for example, or with the Yellow River in China.

Erosion on slopes, or even on plains, will also mean that larger quantities of silt are washed down into the rivers. This may further increase floods, as well as hampering navigation and silting up irrigation works.

Degree of Soil Erosion

☐ Slight or None
▨ Moderate (25–75% of Topsoil Lost)
■ Severe (More Than 75% of Topsoil Lost)

Map 15 Soil erosion in the United States

Normal stream channels become choked with silt, and the river may change its bed. Silt deposition may raise the bed above the level of the surrounding country, as it has in the lower stretches of the Yellow and Mississippi rivers, especially where man has built dikes or levees to prevent the escape of flood waters and the silt they carry. The river itself may build up what are called natural levees, where it drops silt along its banks as flood waters retreat into the normal bed. Successive flood crests may be unable to break out, and silt will be deposited in the normal channel. With the river confined and building up its bed above ground level, the inevitable break-through of pent-up flood waters behind dikes or levees is likely to be disastrous. "Dig the bed deep and keep the banks low" is a sound prescription for river management, devised in ancient China, but it is difficult to apply where *watersheds*[3] have been denuded and rivers are silt-laden.

Changes in effective climate

Quick runoff of rainfall may also change the *effective* climate, since the same amount of rainfall will impart less moisture to the soil. "Effective climate" refers to the usability of rainfall and the extent to which the total precipitation is made available for agriculture or for other purposes. The dust bowl of the 1930s in the Great Plains of the United States was a direct result of the plowing of the less humid parts of the steppe grasslands and the planting of wheat to meet the accumulated world demand during and after the First World War. With the soil thus exposed, a series of dry years weakened the basis of wheat growing and opened the way for winds to pick up the dry dust and bodily carry off millions of tons of soil. The drought and dust storms of Texas and adjacent areas in the 1950s are traceable to a similar set of causes, including the plowing of marginal lands after the Second World War. Steppe and arid-climate rainfall is largely in the form of convectional thundershowers, whose violence accentuates run-off and increases the vital importance of vegetation cover. Tree cover may even have the effect of increasing the proportion of milder showers, because trees decrease heat reflection and radiation from the ground. Removal of tree cover has certainly changed the *effective* climate in every part of the world where it has taken place and has also altered the soil and the rest of the vegetation cover. Animals kept by man have further reduced plant cover in treeless areas with similar unfortunate results, or have prevented the regrowth of trees. It has been well said that "aridity is the work of man" – a pardonable exaggeration which is not as far from the truth as it might seem.

Deforestation

Marginal areas are particularly susceptible to damage, but even in better-watered places where an originally vigorous forest cover has been cut off, erosion and floods may become a tremendous problem. Where the original forest is marginal, it may not be able to re-establish itself by shooting or seeding once it has been cut down, even if cutting is not continued. Forests may persist for some time in areas where the climate has gradually become too dry for new forest growth but not too dry to maintain existing trees. It has been suggested that steppe or a prairie grass type of vegetation now covers many marginal areas which were once forested, because primitive hunters or farmers burned off the original forest to drive out the game or clear the land and set grass fires at regular intervals thereafter for the same purpose. Some students argue that much of the present steppe vegetation of the world, especially in the tropics, may be the result of man burning off an original tree cover.

Diffusion of plants

People have also altered natural vegetation by spreading plants very widely. They have done so as long as they have lived on the earth but especially in the last four hundred years as Europeans established and intensified their contacts with the New World and with Asia and Africa. Wider distribution of useful crop plants like wheat, maize, rice, potatoes, or manioc has basically altered the landscapes, cultures, and economies of areas where these plants were absent before and of other areas related to them. Not only useful or cultivated plants have been spread but many others, usually as "volunteers." There are but few parts of the world where the present supposed "natural" vegetation does not include plants introduced purposely or accidentally by humans, often from another continent. Hybridization, natural and human-directed, further complicates the mixtures. Human modification of vegetation by fire, clearing, plowing, road building, drainage, or other activities also favors some plants (as it favors some animals, such as mice, deer, pheasants, or starlings) at the expense of others. This may fundamentally change the total vegetation complex, apart from the cultivated plants or those which people purposely try to encourage or eliminate. Perhaps the simplest illustration is the profusion of weeds in a plowed field, many of which may be of distant origin, and all of which have multiplied because of human activities. Recent illustrations of the effects of grazing animals on vegetation, apart from the well-known ravages of sheep and goats, were provided by the introduction of the rabbit into Australia, and its near extinction in Britain as a result of a pandemic disease. In both cases the vegetation and the look of the whole landscape were changed.

But society's efforts have been concentrated on the removal of vegetation, especially the forest. People have had an antipathy for the forest ever since the emergence of *Homo sapiens*, and probably before. The forest's dark recesses harbored wild beasts and other unknown terrors, and from a purely practical point of view it was in the way of agriculture and of circulation. People have only recently, after many unfortunate experiences, learned that in some cases retention or careful modification of the natural vegetation may be better for them in the long run than its removal.

NOTES

1. Vernon Bartlett in the *Manchester Guardian Weekly*, Nov. 22, 1956, p. 15.
2. Especially in Brazil, plantation and mining developments have invaded and cut over large areas of the Amazon Basin in recent years, a trend still continuing at an alarming rate. There is growing concern among earth scientists worldwide about the ecological consequences of continued removal of the rain forests.
3. The drainage areas of rivers and tributaries: their source of water from rainfall and run-off.

10 SOILS AND AGRICULTURE

Agriculture is still the most indispensable productive activity, and soils are therefore a primary resource, more important than coal or iron ore or petroleum. Like mineral deposits, soil can be exhausted, although broadly speaking soils may be a product of climate in a more direct or current sense than is the case with coal or petroleum. It still takes thousands of years, however, even if not the millions needed for coal, for a mature developed soil to form. Since on such a time scale climate may change considerably, the pattern of soils and their distribution is complex. Most present soils are probably in large part "fossil" soils derived from an earlier glacial or preglacial period and also acted upon by subsequent climates. There are factors other than climate which influence soils, which are discussed below. Nevertheless, the general outlines of the major soil groups of the world coincide roughly with climates in their distribution, as can be seen by comparing the two world maps (pages 91 and 129). It was the Russians who first realized in systematic fashion a causal relationship between climates and soils; a careful study of the two maps will show that the Russians had certain advantages in developing this idea.

A number of new soil classification systems have been developed, as the result of research during the past couple of decades, which attempt especially to reflect soil formation, like all other environmental elements, as a dynamic process and which also allow for the great range of factors which affect this process. Soil classification systems are complex, and involve technicalities and details which need not concern us here, since our focus in on human *use* (and misuse) of the physical world. However, it should be recognized that the discussion of soil types here is highly general and that a great deal more information is readily available, increasingly now from the work of the Soil Survey Staff of the Soil Conservation Service of the United States Government. This agency has published a classification system titled (appropriately enough) "The

Map 16 World soil types. *This is necessarily highly generalized so as to make it intelligible at this scale. However, compare this pattern with that of climates (see Map 11, page 91).*

OXISOLS
Nutrient-poor, high in iron and aluminum. Deeply weathered.

SPODOSOLS
Accumulation of amorphous material under light, sandy layer.

ULTISOLS
Low organic matter, low base supply, layers of clay accumulation.

VERTISOLS
High clay content of swelling clays. Deep cracks in dry periods.

WORLD SOIL TYPES

SCALE

GOODE HOMOLOSINE EQUAL-AREA PROJECTION

MOUNTAIN SOILS
Temperature and rainfall vary. Variable slopes and, elevations.

Little or no soil

ALFISOLS
Podzolic; gray to brown surface horizons. High base supply.

ARIDISOLS
Dry for more than 6 months of the year.

ENTISOLS, INCEPTISOLS
Weak or absent horizon development.

HISTOSOLS
Mucks, bogs, organic soils.

MOLLISOLS
Black, organic-rich surface horizons. High base supply.

Seventh Approximation." A good clear discussion of this system, and of the technical properties of soils and their classification, can be found in A. H. and A. N. Strahler, *Elements of Physical Geography* (Wiley, 1979), Chapters 10 and 11, or in *Modern Physical Geography* by the same authors (1978).

Soils and climate

Climate may allow or prevent the retention or accumulation in the soil of soluble mineral elements and of *humus* (decayed organic matter). Most minerals in the soil are soluble, except silica, iron, and aluminum oxides; where rainfall is heavy and continuous and temperatures are high so that oxidation is rapid, most humus and minerals, except silica, iron, and aluminum oxides, are removed. *Leaching* (removal of soluble elements by water) and oxidation, when carried to extremes, produce a material which is not only infertile but may be literally as hard as a rock. Such soils are frequently called *laterites*, from the Latin word *later*, meaning brick, because they may be as hard as a brick and are often given a brick-red color by insoluble oxidized iron. By no means all red soils are laterites, nor are all laterites red, since redness is simply the result of the presence of oxidized iron, which occurs in a great many soils in warmer climates where oxidation is relatively rapid. Laterites are especially thoroughly leached and oxidized. In many parts of the tropics where laterites occur they are actually quarried and used as building or paving stones. Leaching and accumulation of iron and aluminum also occurs in the grey, acid soils of humid but cooler climates, where such soils are frequently labelled *podzols*. It has recently been demonstrated that podzolization and laterization are in effect the same process, resulting in essentially the same end product, even though the color, texture, and compactness of the soils may differ as a result of lower oxidation in cooler climates.

Leached soils are generally of low fertility and are particularly deficient in the highly soluble mineral nutrients which help to produce protein in the crop. Such soils may be able to produce good yields of bulk carbohydrates, but protein content will be low, and the crop will therefore be less valuable as food or as animal feed. This is a common problem in the wet tropics and has also been analyzed in some detail by comparing humid and arid areas in the United States. The luxuriant forests which occur on many tropical lateritic and cold-climate podzolic soils, benefiting from plentiful moisture or high constant temperatures, give a misleading impression of soil fertility. Such soils cannot usually support annual field crops for more than a short time before their slight fertility is exhausted, since most annual crops are more soil-demanding than trees.

The effectiveness of a given amount of precipitation depends in part on the evaporation rate, as discussed in Chapter 8. Ten inches of rainfall in northern Canada with low evaporation may produce a highly leached soil, whereas ten inches in southwest United States with high evaporation will not. Where the effective rainfall is slight, more of the soluble mineral elements are retained and the soil is likely to be much more fertile. High constant temperatures also speed up the decomposition of humus, or dead vegetation, and its complete oxidation and leaching. Low temperatures retard decomposition and oxidation and may lead to acid soils or soils poor in humus content, since the humus merely lies around as surface litter and generates acids. Podzolic soils are formed under these conditions, whereas laterites are formed where high constant temperatures allow little or no accumulation of humus. True laterites are found only in the equatorial climates, and true podzols only in the humid, cold climates. Between these extremes and the desert regions there is a complex gradation of soil types which broadly reflects the accompanying gradation of climates. As usual, the extremes are unfavorable: in the equatorial tropics, laterites, in the cold areas, podzols. In the desert, the deficiency of rain may mean the absence of humus and the accumulation of salts or alkalis which are not washed away as they are from the soils in humid climates and which poison the soil for plants.

Humus

Vegetation plays a major role in the development of soils, by providing humus and by protecting the surface layers. Humus is probably the most important positive factor in soil building. The most fertile of all soils lie in the general zone of the steppe, where decaying humus from deep-reaching grass roots has built up a deep, dark, rich layer of earth. Humus is important for the structural qualities which it gives to a soil, promoting good circulation and retention of air and water, and for the rich plant nutrients it provides. Humus is also an essential aid in the process of biological and bacterial action which helps to keep a soil aerated. Good soil is alive in the sense that it supports a myriad of bacteria, earthworms, and other creatures. Their work of decomposition and circulation and their own dead bodies are of tremendous benefit to plants. This live matter is often present in a soil in direct proportion to the amount of humus which the soil contains. Where vegetation is absent or sparse, or where vegetation decomposes too rapidly or too slowly, the productivity of the soil suffers seriously, apart from the soil erosion which may also accompany the absence of vegetation. Some types of humus are better for most crop plants than other types. For example, coniferous or needle-bearing trees yield a humus which is inclined to be acid; where such trees are the principal suppliers of humus, only plants which thrive in or tolerate an acid soil will prosper. Such plants include the conifers themselves, the rhododendron family, pin oaks, blueberries, hollies, and some lilies, but very few of the major commercial or food crops.

Parent rock

What is called the "parent material" also affects the nature of the soil. The soil's original content of minerals is inherited from the parent rock, and soil structure or texture may also be derived from the parent materials. Heavy clay soils may be the result of weathering of one type of rock, sandy soils of another. Soils formed, for example, in regions of limestone rock are often likely to be fertile, those in regions of granitic rock less so. But climate exercises an over-all influence, leaching away the fertility of an originally well-provided soil if it happens to lie in an area of heavy rainfall, or allowing the accumulation of humus and retention of minerals in a granitic soil if it lies where rainfall is moderate or slight. Small areas in every part of the world reflect in their soils the parent rock. Europe and the United States, for example, are dotted with fertile limestone soil areas which contrast sharply with surrounding less fertile areas and with regions of heavy clay soils where poor drainage affects natural vegetation and cultivated crops. In most of Europe and America rainfall and temperatures are moderate enough to allow fertility to be maintained, at least to a degree and for a time. But even in the rainy equatorial tropics local soil distinctions based on parent rock are apparent.

A somewhat special case is soils formed from the weathering of volcanic lava or ash deposits. Such soils are often deep enough to resist leaching, and they may be frequently renewed by periodic lava or ash deposition, so that climate affects their fertility relatively little. Where their chemical composition is basic rather than acidic, they may be very productive agriculturally. The island of Java has already been cited in

Figure 20 **Sample soil profile.** *Typical* Chernozem *soil profile. This steppe soil forms in semi-arid climates and hence accumulates sub-surface humus and soluble minerals such as lime. However, most soils display a more or less clear profile composed of distinct* A, B, *and* C *horizons.*

Chapter 9 as an example, and there are many other similar areas of fertile volcanic soil, such as parts of Cuba, the Columbia plateau in north-western United States, and parts of the Deccan plateau in south India. In broader patterns, however, the role of climate is clear. From the same original materials in two different climatic situations, two quite different soils will tend to develop (see Figure 20).

Transported soils

Some of the most productive soils are the result of erosion and subsequent redeposition in another area, by water, wind, or ice. The materials which rivers transport and redeposit along their courses as *alluvium* stand out on the world map of soils as they should, for they form among the most fertile of all soils. If it flows through a variety of rock and soil areas, a river picks up a great mixture of sediments. River silt is thus likely to be well balanced chemically and to contain most of the mineral elements neces-sary for plant growth. The usually deep soil formed by river deposition is also likely to have good drainage and structure and a fine texture, depend-ing on the velocity of the river, for a stream can carry larger particles or even rocks in proportion to the force of its flow. As the velocity decreases, the larger particles are dropped, and the finest alluvium is found in the lower courses of rivers or in their *deltas* where, after reaching a relatively flat plain, the rivers are flowing more slowly. Areas of alluvial soil, as these riverine deposits are called, are particularly valuable when they occur where neighboring soils are poor, as in much of the tropics. This is an important reason for the remarkable concentration of the population of India, China, and peninsular southeast Asia in river valleys. But alluvium nearly everywhere in the world is at a premium. *Loess* is akin to alluvium, but is believed to have been laid primarily by wind rather than by water. Strong out-blowing winds from the retreating glaciers at the end of the last ice age may have picked up finer particles from the glacial out-wash plain, deposited earlier by melt water from the ice, and distributed this dust in thick layers on top of extensive areas to leeward. Loess areas include much of north and northwest China, parts of northern Europe, and parts of central United States. In fertility and structure, loess is similar to river alluvium. Many large areas are also covered with soil originally dumped there by advancing or retreating glaciers, or by the gradual shrinking of lakes formed by the melting of ice sheets. Some of the glacial deposition may be stones, gravel, or sand, as in the case of materials deposited by swift-flowing rivers, but river alluvium may also be matched by fine-textured and well-balanced glacial deposition of varied origin which is highly productive agriculturally, for example, large areas of the northern half of the American Middle West.

Soil productivity

There is no need to be concerned here with the technical aspects of soils and soil types. Actually, even the soils scientists are increasingly reluctant to distinguish general soil types on a world-wide basis and are not sure how important color, texture, levels, and other distinguishing marks are to the agricultural properties of a soil. Soils are of interest here only as they may help in understanding people and their works. The major question is whether a given soil is actually or potentially productive, or whether it may have special characteristics which may or may not fit it for specific uses. Fertility is one of these characteristics and one measure of a soil's productivity. Fertility in its strict sense means the available mineral content of a soil and in particular its content of adequate soluble amounts in the right proportions of the main elements essential to plant growth: nitrogen, phosphorus, potassium, calcium, and others in lesser amounts. The structure, texture, and depth of a soil are also of great importance, since plants depend on a circulation of air, water, and nutrients in solution to their roots. A densely packed soil or one with either poor drainage or with so much drainage that the water percolates right out of it will obviously not have proper circulation. However fertile it may be in its mineral components, it will not be very productive. Soil which is too shallow also hampers proper plant growth; it quickly dries out, and drainage is likely to be poor. Some soils can be plowed more easily and effectively than others. Some retain good circulation of air and water after plowing, while others cake together with the first rainstorm or bake hard with the first dry spell.

As a rule, lack of fertility is easier and cheaper to remedy than unfavorable structure, texture, or depth. Chemical or organic fertilizers can be applied, but the technical means of correcting soil structure are more limited. In both cases, however, it is economic rather than technical considerations which restrict the artificial improvement of a soil. It costs money to fertilize, to dig in humus, to lay drains, or to apply chemical soil conditioners. It pays to spend this money only when an appropriate return can be realized. In practice, such improvements are generally limited to the favorable climatic areas, to irrigated areas, or to areas characterized by intensive land use because of population pressure or because of market considerations. It often may not pay to fertilize, for example, the huge wheat fields of semi-arid, thinly settled, and relatively remote western United States, Canada, or Australia. In their marginal climates the cost of fertilization would not be repaid by profitable increases in wheat yields. It does pay to fertilize, and if necessary to lay drains, in the irrigated Imperial Valley of California, the truck-garden areas of New Jersey, or the rice-growing deltas of the Orient. In each case location or local conditions are such that soil improvements are worthwhile.

Human treatment of the soil

Fallowing and rotation

People modify the soil in many ways. Fertilization with chemicals, pulverized rock, manure (which is rich in nitrogen), or humus is only one way, and not necessarily the most important. Depletion of the soil may be slowed by *fallowing*, or allowing the soil to rest for a shorter or longer period between crops. Where the soil is naturally not very fertile or where the crop being grown is especially soil-exhausting, fallowing is often essential. It is characteristic of the agriculture carried on in large areas of the tropics outside the alluvial or volcanic soil regions, where heavy rainfall aided by strong topographic relief and by high constant temperatures has badly leached the soil. Farming in these areas is often *shifting agriculture*. A given plot may be cultivated for only one, two, or three years before its soil is exhausted and the farmers move on to clear another plot. They will return to the first plot only after several years of fallow, the period varying with the type of soil, climate, relief, and crops grown, and also with the pressure of population. On each small plot in turn, the forest is cleared by burning, girdling, or felling, and the litter is burned. The ash provides a quick fertilizer. When the plot is abandoned, the second-growth jungle in a tropical climate quickly recovers the land so that it is protected from erosion. Shifting cultivation is often called a wasteful form of land use. It progressively or continually removes the natural forest cover or second growth, and any one farm family must range over a very wide area, from which they can produce proportionately little. But it is difficult to devise a less wasteful or more productive agricultural technique for such areas, especially since the technical or economic advantages of a commercial economy are largely lacking. Fallowing is practiced in most agricultural systems to some degree, and in drier areas it also has the advantage of conserving the moisture in the soil, which is of course drawn on heavily by any crop. In the European Middle Ages, fertilization was not well understood or heavily used, and fallow occupied one-third of the "three-field system" in order to restrict soil exhaustion.

Another widely practiced method of conserving soil fertility is the growing of crops in *rotation*. Each crop takes a different combination of nutrients from the soil. If the same crop is grown in the same field repeatedly, yields tend to decline. But if the same field can grow a variety of crops in successive years, the yields of each can be better maintained and the soil balance better preserved. Some crops actually add nutrients to the soil, most importantly the *legumes*, which include peas, beans, clover, peanuts, lentils, and alfalfa. Leguminous plants extract nitrogen from the air, which then forms in little globules on their roots. They leave a soil more fertile, especially for heavy nitrogen-demanding crops like most cereal grains, than it was before the legumes were planted. Legumes are therefore often used in crop rotation, alternating with grains.

Other crop plants may help to aerate the soil by their root development, especially if they are uprooted when harvested, sugar beets, for example. Still others help to hold the soil in place and protect it against erosion by water or wind, whereas some are not good *cover crops* and invite erosion, especially on slopes. Most row crops are not good cover crops, since they leave much soil exposed between the rows. Cotton, maize, and tobacco are particularly conducive to erosion, both lateral and vertical, since rain percolates through exposed soil more rapidly. Cotton and tobacco also make especially heavy demands on soil nutrients. Since each crop has its own particular soil demands or soil preferences, an intelligent farmer, if he has adequate choice, must use care in combining and rotating crops and soils, especially if he wants to preserve his soils, as well as maximize his yields. Most farms grow more than one crop, in part for this reason, since even within a small area soils, drainage and soil moisture, slope, and often even insolation and air drainage (important for minimizing frost damage) will vary. The growing of different crops also helps to spread labor needs more evenly over the agricultural year, restricts the spread of plant diseases and pests, and minimizes financial risks due to failure of any one crop or to varying market prices.

Cultivation and weeding

Agriculture also alters the soil by cultivating it. Strictly speaking, cultivation means the stirring or aerating of soils by plowing, digging, scratching, raking, or sometimes just by making holes in it with a pointed stick or other instrument. All these techniques help air to circulate in the soil and also break the soil up into finer particles so that water and nutrients in solution can circulate better. Cultivation may also improve drainage, at least in the upper layers. Although it may increase evaporation of water from the soil, cultivation can also help to reduce surface evaporation if it pulverizes the surface layer finely enough and often enough to break up the capillary action by which water moves to the surface. Frequent harrowing, or pulverizing of the surface layers, is commonly done in dry areas to conserve soil moisture, especially during fallow periods when an additional surface layer of humus, or even of crushed stones or pebbles in areas of sparse vegetation, may be applied as a *mulch* to restrict evaporation.

Plowing can accelerate erosion, especially if the furrows run in the same direction as the slope, or can significantly decrease erosion if the furrows follow the contour of the land at right angles to the slope. *Contour plowing* has saved many farming areas from destruction and can help to safeguard almost any cultivated land. Even a slight slope hardly perceptible to the casual eye may generate serious erosion if it is left bare in a climate of heavy rains, and on perfectly flat land heavy rains may produce *splash erosion*, especially if the vegetation cover is inadequate. Some soils, the lighter or sandier ones, erode more easily than other heavier or clayey soils.

Most soils in areas of moderate or light rainfall profit from frequent cultivation. Usually, the deeper the plowing, the better the soil retains its productivity, not only because aeration and circulation are improved, but because soil layers beyond the reach of most annual plant roots are turned over by a deep plow and brought to the surface or to the root zone where plants can use their less exhausted nutrients; this depends, of course, on how far below the surface the soil remains fertile. The soil also profits if some of the annual crop can be left behind after harvest – roots, stalks, leaves, or whatever is not going to be consumed – to form humus and to replace some of the minerals which the crop has taken from the soil. Agriculture breaks the natural cycle of plant growth and decay by removing the plant, or part of it, when it is mature so that its decayed materials are not returned to the soil. Unless the nutrients can somehow be replaced, the soil suffers. Agriculture also interferes with the natural plant and soil balance by allowing only one crop plant in any one field instead of the natural profusion of plant communities and thus puts a heavy strain on the soil. Weeds take nutrients and water from the soil, too, and leave proportionately less for the desired crop. If weeds can be left in the field to rot when they are cut or pulled up, the soil also profits and, with the dead weeds as a mulch, the soil will have extra surface protection against evaporation and erosion.

These generalizations about cultivation and weeding apply well enough to soils and agriculture outside the tropics. But where rainfall is high in amount and violent in character, or where temperatures are constantly high, leaching and erosion are accelerated by the clean weeding of a field and by the plowing of the soil. Traditional farming practices in many parts of the wet tropics try to leave as much vegetation cover as possible on the soil, to weed as little as possible, to avoid row crops, and to plow as little as possible or to rely on aerating the soil with a pointed stick. Crops may be grown crowded or jumbled together, and the field may look most disorderly and poorly cared for to an outsider. It is only recently that technical advisers from outside the tropics have begun to realize that such practices make good sense.

Agriculture and population pressure

Population pressure exercises a powerful influence on land use, and indeed one can analyze all agricultural systems and distinguish between them in terms of population pressure. Where few cultivable acres must support many people, as for example in much of monsoon Asia, most farmers cannot afford to leave the roots or stalks of their crops in the field; crop residues are needed for fuel or for animal fodder. Few Asian farmers can afford to apply adequate amounts of manure, for the agricultural system cannot support enough animals, and manure is also in demand as fuel. Few can accord to fallow their fields or to rotate crops; since the farmers must obtain maximum returns per acre every year, they must devote their land to growing the one most productive crop year after year or season after season.

Fortunately it is rice, grown for the most part in standing water, which dominates the agricultural system of much of monsoon Asia, and rice is in many ways a unique crop. It takes nutrients from the soil, but they are largely if not entirely replaced by nutrients which enter with the irrigation water or are derived from animal and vegetable life in the irrigated paddies, especially the blue-green algae which thrive in the high temperatures present and which are potent fixers of nitrogen from the air. The water in the paddies also protects the soil from the damaging effects of wind and rainfall erosion and leaching as well as from evaporation. Other nutrients are added by the farmer, including night soil (human manure). Weeding and cultivating are, however, assiduously practiced, methods are in general highly *intensive* (much labor applied to relatively small amounts of land), and crop yields per acre are very high. In the less densely populated parts of the world fewer people have to be fed from each cultivated acre, the farming can be more *extensive*. Where income from manufacturing and trade is relatively high and agriculture is not the only economic support, good farming practice often plows under a crop of legumes or of something else in order to enrich the soil without taking away any of the crop. This is called *green manuring*, and it is fine for the soil, but clearly the peasant farmer in monsoon Asia cannot afford to do it. Population pressure also helps to explain shifting cultivation and the changeover to sedentary agriculture when it occurs, irrespective of soils. As population mounts in any area, agriculture must become more intensive and produce higher per-acre yields.

Tree crops

Perennial or tree crops are usually not as sensitive to soil conditions as are annuals, which must in effect produce an entire plant from seed in one growing season on the basis of what the soil provides. Most tree or perennial bush crops do need good drainage, and since their roots go deeper than most annuals, orchards and plantations must often pick sites and soils with some care, avoiding those which are poorly drained. Fertility may be less important, and poor soil areas may be well forested or may support good tree crops. Orchards often make use of slopes, where drainage is likely to be good but where soils are often poorer than on level land because they are more subject to erosion and to soil slip and slide. Trees help to hold the soil on slopes, and fruit trees may benefit on slopes by avoiding the lower pockets of land where cold air tends to collect and frost damage is higher; fruit blossoms are especially susceptible to frost damage. In the tropics, where most plantation agriculture is located and where most soils tend to be poor, tree crops like rubber, tea, oil palm, coconuts, or cacao (chocolate) can yield well on lateritic soils which could not profitably support annuals. Trees do, however, reflect soils, like all vegetation. Outside the tropics, coniferous trees are usually more tolerant of poor soils (and of cold climates) than are the hardwood or deciduous trees.

People's effect on the soil

Whether they improve the soil or degrade it, people alter it whenever they use it for any purpose. Chapter 9 has already examined some of the effects on the soil which modification of the natural vegetation may have. If only out of immediate self-interest, most farming tries to improve the soil, or at least to make it remain productive. Good farming methods can transform an unproductive soil into a highly productive one, if the farmer has the technical means and the economic ability to invest labor and capital in the soil. In balance, most long-established farming is of this positive and constructive sort rather than the destructive farming which is often called "soil mining," because it takes from the soil until fertility is exhausted and does not contribute proportionately to the land. Such practices, apart from shifting agriculture, may mean that the land will have to be abandoned after a brief period of heavy use. Too much of American agriculture has unfortunately been of this "soil-mining" type. Cotton and tobacco, maize and wheat, all important in the United States, are soil-exhausting crops which have often been grown for short-term gain. Application of chemical fertilizers alone seldom restores the soil balance adequately. The agricultural system in current use may often be the major factor in determining what the soil can physically support or what the soil's physical properties are, since all farming transforms soils, for good or bad.

Irrigation

A separate problem is associated with irrigation, especially in dry and semi-arid climates where of course irrigation is most needed. Most water contains minerals in solution, including salts and alkalis. When irrigation water evaporates, as it does rapidly in hot dry climates, the minerals are left behind on the surface. Salt and alkalis are poisonous to plants, especially to most crop plants. Where they are not regularly flushed or diluted by rain or snow they may build up on the surface and poison the soil. Sooner or later this happens in nearly all irrigated areas in hot dry climates. Large areas in places such as the American southwest, northwest India, or northwest China, have been plagued by this problem, and much cultivated land has had to be abandoned. Mineral accumulations can be flushed out with pure water, but this is often scarce and expensive. The end result of mineral accumulation can be observed in extreme form in the salt and alkali flats in many desert areas, such as the Utah Salt Flats around Great Salt Lake.

Economic and social influence of soils

Soils may often be reflected in the social or economic structure of an area. A good soil area may be a relatively prosperous one, not only for agriculture but for the many service industries, market centers, and large population which a rich farming area may support. The productivity of Iowa farms contributes to the size and manifold functions of the city of

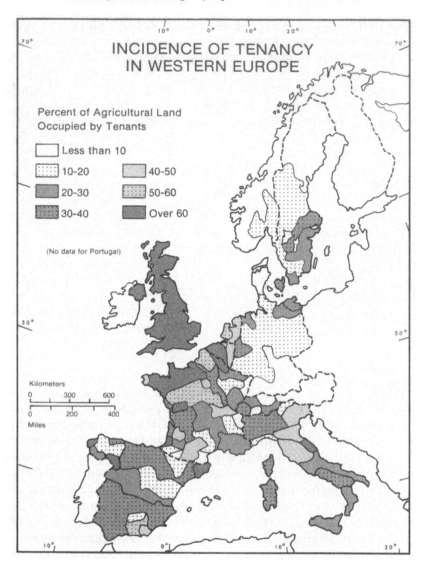

INCIDENCE OF TENANCY
IN WESTERN EUROPE

Percent of Agricultural Land
Occupied by Tenants

☐ Less than 10

▦ 10-20   ▦ 40-50

▦ 20-30   ▦ 50-60

▦ 30-40   ▦ Over 60

(No data for Portugal)

Kilometers
0   300   600

0   200   400
Miles

Map 17 Tenancy in Europe

Chicago, which is primarily a service, trade, and manufacturing center for
the agricultural Middle West. Soils and climate play their part also in the
size of New York City, and indeed of the whole national economy, which
in differing degrees finds markets and services to perform in or for the
populous and prosperous agricultural areas. As a primary resource, soils
are an important item of national strength. They may permit considerable
self-sufficiency or surpluses for export in agricultural products, or they
may prevent surpluses and help to require expensive or strategically
awkward imports. Regional or national soil differences, like other
regional differences in environment, may help to stimulate complementary exchange between regions or nations.

A good soil may also attract heavy investment in the form of irrigation or in the form of specialized plantation or truck-garden agriculture. A poor soil area is likely to be much less prosperous as a whole, unless there are compensating advantages in other respects, and its low economic status may be reflected in the size and condition of its houses, its roads, its schools, and perhaps its political sympathies. In middle-western United States, where some study has been made of the correlation between soil types and voting habits, there has been considerable coincidence between conservative politics and good soils on the one hand, and between liberal or radical politics and poor soils on the other. If soils are very good, they may be reflected in a high price for land and may therefore in some cases be cultivated largely by tenants who cannot afford to buy the land but who pay a high rent for it. The land may be very productive without returning a high average income to the individual cultivators, or it may support a reasonably prosperous tenantry. The map of tenancy in western Europe (Map 17) shows a clear relationship between high tenancy and the combination of fertile soil and favorable climate when the spatial distribution of tenancy is compared with the spatial distribution of soils and climate.

In other cases good soils may attract a dense population into a semi-arid or fluctuating climate, as in north China, where the shortage of rainfall contributes to the fertility of the soils. Recurrent famines often resulted because there were limited means of regional exchange to supply the area with food when rainfall was inadequate and crops failed. Good soil may sometimes join with favorable climate to invite what may be called "fussy" crops, which are not tolerant of low fertility or poor soil structure and are climatically demanding. Cotton, tobacco, sugar-cane, indigo, and jute are fussy in this sense. Land which can grow them is likely to be expensive, and these crops also require large amounts of hand labor (in some cases recently displaced to some extent by machinery). In many areas where these crops have been grown, slavery, tenancy, or contract labor has been the solution applied.

Soils and the American Civil War

Maps of soils and climate both show a break in their patterns for eastern United States which coincides almost exactly with the line between Confederate and Union territory or sympathies during the Civil War. What is now West Virginia, still called by some of its inhabitants "West by God Virginia," lies outside the soil and climate patterns which cover most of the Old South. As a largely mountain or upland area, West Virginia's relatively poor soils and cooler climate could not support the plantation agriculture worked by slaves in lowland Virginia and in most of the rest of the Confederate south. South of West Virginia the Appalachians carry these relatively poor upland soils, and there, too, in the mountains of Tennessee and the Carolinas, Confederate sympathies were often lacking. Most of Kentucky, the poor-soil upland parts of it, belonged in the northern camp politically. One of the strongholds of plantation agriculture in the south was and is the alluvial-floored valley of the Mississippi, a logical center for the institution of slavery. Slavery was the root and body of the conflict between north and south, and it was clearly related to these broad differences in soil and climate between the two combatants, although there were, needless to say, many other factors than soil and climate differences which accounted for slavery. Similar distinctions between types of social and economic organization, though seldom with the same tragic consequences, can be made in part on the basis of soil and climate differences in many other parts of the world.

Soils and urban sites

In special cases, local soils may affect urban siting, since this involves questions of drainage, foundation for buildings, and vulnerability to earthquakes. Recently deposited alluvium may be avoided for large buildings in areas where earthquakes are frequent. Uneven settling and foundation problems in some soils, especially alluvium, may make special and expensive building techniques necessary. Soils with poor drainage pose problems for sewage, roads, and the maintenance of structures. Usually, such local site problems are less important than wider considerations of location and access. In fact, many cities have arisen on poor local sites because there was no preferable alternative without sacrificing access. All of the site problems mentioned above, and many others, are common to most river deltas, for example, and yet many of the world's largest cities have grown up on deltaic alluvium. Location near the mouth of a navigable river is advantageous enough to outweigh problems of drainage or building. New Orleans is a classic example; much of the city actually lies below the level of the Mississippi River, which at New Orleans is confined behind natural and artificial levees and has built up its bed by silting. One of the problems of New Orleans is sewage disposal, since wastes cannot be dispersed by gravity and have to be pumped up to the river. There are many other deltaic cities in other parts of the world which have grown despite local soil or other site problems. Freetown in Sierra Leone has a different sort of problem. The vicinity of the city is covered by a particularly hard and impermeable laterite crust which has discouraged even natural vegetation except for some sparse grasses. Freetown has grown up in a small "island" where there is a surface layer of workable soil over the laterite crust.

Intensive and extensive agriculture

Soils play a part in determining whether an agricultural system will be intensive or extensive. Intensive cultivation is more likely to be profitable on a good soil than on a less good one, if it is combined with a favorable climate or if it can be irrigated. A poor soil, or a soil in a less favorable climate, is more likely to be cultivated extensively since it is less likely to return an adequate recompense or yield to repay intensive investment.

There are many exceptions to this. The fine steppe soils are for the most part extensively cultivated, and the much poorer soils of densely settled parts of the tropics or subtropics are worked intensively. Climate is largely responsible for these exceptions, but it does not entirely destroy the association of soils with the degree of agricultural intensiveness. Poor soils are worked intensively, as noted above, on tropical plantations which specialize in tree crops. The availability of capital and the pressure of population are clearly other factors which will affect the intensiveness of soil use. But soils are one factor which helps to determine whether investments in any form will pay.

What determines soil use

The uses to which a given soil may be put will depend for the most part on the following factors:

What the soil will physically support

The soil demands or soil tolerances of different crops vary widely. "Fussy" crops have particularly stringent requirements. Others are tolerant of poor soils, or of soils with particular characteristics. These less demanding crops would include, in addition to trees, potatoes, rye, millet, and hay. Where they are grown they may provide some evidence of poor soils, although climate may be even more important in accounting for their presence. Dairying often tends to occupy poorer soils because it does not need good ones (for pasture) as much as field crops need good soils and is therefore less willing to pay for them.

The nearness, nature, and size of the market

Poultry raising and dairying are especially sensitive to market location, because their products are perishable, bulky, or present proportionately expensive transport problems. But the best soil in the world is of little use for any crop without an economically accessible outlet for its products. The location pattern in the dairy industry of cheese, butter, and fluid milk production at different distances from the market has already been discussed in Chapter 5. Most large cities, at least in Europe and North America where milk is important in the diet, support fluid milk production immediately around them, whether the soil and climate are suitable for dairying or not and even if fodder must be brought in from other areas. Poultry, especially eggs, are also dependent on quick access to market, and the industry tends to cluster in the vicinity of large cities, sometimes even in their suburbs. Generally, the farther a soil is from the market, the more extensively it will be used, although it may often be devoted to high unit value crops or to crops which are easily and cheaply shipped long distances. Wheat is a good example. It keeps well, ships easily, and can be loaded and unloaded mechanically. Most commercially grown wheat comes from the steppe areas of the world, which are remote from major population centers. Wheat is more tolerant of dry climates (though not of poor soils) than most crops. In the steppe areas far from markets it is grown extensively. In the more humid areas close to markets, such as northwest Europe, it is grown intensively. In the United States, wheat gives way to maize (corn) eastward as the climate becomes moister and the growing season lengthens.

Corn is a fussier crop than wheat, and where climate allows, it is more profitable. In the eastern part of the corn belt, near large market centers, corn is shipped in bulk to the market since it does not have far to go. Farther west, more of the corn is fed to cattle or hogs, thereby converting it into a higher unit value product which can better bear the cost of long shipment. This choice of alternatives is also affected by the relative prices offered in the market for corn and hogs or beef, which continually vary. The corn belt farmer, especially if he is in a median location close to the break-even point between shipping corn and shipping hogs or cattle, has to keep in mind what is called the *corn-hog ratio*. At certain times market prices may make it more profitable for him to ship his produce as hogs, at other times as corn. Specialization of this kind, or of most kinds, is not possible in areas which lack cheap transport. Soil or land in such areas, despite its regional variations, is more uniformly used to grow cereal and other high-calorie food crops for immediate local consumption by dense subsistence populations. Market distances are therefore less relevant. In commercial economies land near the market will generally have a high rent because of competing uses and because market nearness minimizes transport costs and maximizes net profits. Intensive use of the land is therefore more necessary and is in keeping with the principle that a scarce or high-cost resource or production factor will be used intensively. Thus there is a pattern of increasingly intensive use of the land in commercial economies as the market is approached.

The cost of transportation

Distance may be measured in miles, *real distance*, or in costs, *economic distance*. Economic distance is certainly the more accurate and important measure in analyzing the nature and distribution of most of man's activities or in determining the uses to which a given soil may be put. Cost more than miles gauges market accessibility. Transport costs depend on a variety of factors. The volume, constancy, and direction of the flow, the type and capacity of the carrier, and the nature of the route are fully as important as the real distance.

The competition of other areas

A soil or an area may be well suited to a particular use, but another elsewhere may be better, or the agricultural system in another area may produce at lower cost due to more extensive methods, local advantages, or more efficient operation. Still another area may be nearer to the market, in miles or in economic distance. This competitive relationship is likely to be changeable, especially as transport costs change. Commercial wheat growing in the steppe, for example, was not possible until cheap rail and sea transport was developed. Extensive wheat-growing methods in the steppe, developed in the nineteenth century and dependent on mechanical cultivation, made possible a low unit cost of production despite the marginal climate and the relatively low yields per acre. The agricultural development of the steppe and of the American Middle West were partly responsible for the relative decline of agriculture in old England and for the widespread abandonment of farms in New England beginning about 1830. Much of New England is now essentially a series of clearings in a secondary forest where once were cultivated fields whose old stone wall borders punctuate almost any walk in New England woods. Much of the Connecticut valley, however, is still cultivated profitably, thanks to fertile alluvial soil and nearness to markets which combine to keep Connecticut valley agriculture competitive. European agriculture has undergone similar changes as technological development and transport improvements made overseas agricultural production competitive on the European market.

The available technical means

Some soils require special techniques and cannot be used, or their use will be restricted, until these techniques are available. Much of the Great Plains could not be cultivated profitably until stronger plows and tractors to break the tough sods and mechanical harvesters for extensive farming of large fields were developed. Mechanical cotton pickers helped to push cotton growing onto the soils of California at the expense of fruit crops. New crops, introduced from other areas or developed to tolerate certain conditions, may make previously neglected soils usable. Irrigation and drainage can revolutionize the agricultural potential of a soil. The availability of particular techniques varies widely. In pre-industrial areas, technical limitations help to restrict the variety of possible soil uses.

Land near the market

Where market considerations are dominant, special soil characteristics may become important. The middle and south Atlantic coast of the United States, for example, is generally not a particularly fertile soil area, but its sandy soils are mainly light in texture, well drained, easily tilled and fertilized, and they warm up early in the spring, especially since they lie near the sea. Their nearness to the large market centers of the east coast makes it profitable to use these soils for perishable truck crops, which find good drainage, fine texture, and early warming especially advantageous qualities in soil. Truck crops from the sandy coastal plain mature earlier in the year and command a premium price on the market when the heavier clay soils have not yet produced a crop. Denmark and parts of Holland, with soils similar to the Atlantic coastal plain in the United States and similarly located with respect to large market centers, have similar agricultural specialties. In the United States, with its highly developed system of fast low-cost transport, country-wide competition from fruit and vegetable crops grown in Florida and California limits the scope of market gardening around cities in other parts of the country. Local market gardens, however, have a net advantage in summer, when they can put many perishables on the local market at a lower cost and in a fresher condition. Market garden areas are distinguished from truck gardens or truck farming because they are right next to the city, whereas truck farming may be carried on wherever transport gives it access to a market at bearable cost. Market gardens have to use what soil they find in the city environs, but even if the soil is poor, nearness to the market outweighs most soil disadvantages and makes it profitable to invest in soil improvements. Chance pockets of good soil around cities will often attract market gardening or horticulture and greenhouses. These specialties are agriculture of very high value, and the good soil pockets may thus escape the outward spread of housing considerably longer than surrounding suburban areas with poorer soils. When the price of land reaches a critically high point as the city expands, agriculture yields to housing, retailing, manufacturing, or other non-agricultural land use which will pay a higher price for land than farming can pay.

With soils, as with the environment as a whole, physical factors do not operate alone and may often be of relatively minor importance. The significance of location and of spatial relations especially is clear in accounting for land use and may often override the importance of the local soils. Technological or demographic (population) factors may also be of controlling importance. But the environmental influence on people is seldom limited to the immediate physical or economic opportunity which it offers or denies to them. The ramifications of a particular agricultural system and its requirements, or of the general economic level to which a soil may contribute, will often go far beyond a simple physical or local relationship to involve widespread aspects of the society as a whole. This is generally true of the overall relationship between people and their environment.

11 WATER AND PEOPLE

Of all the components of the environment, water is the most crucial, since life cannot be sustained without it, including plant and animal life. Water is also limited, both in total and in each area, to varying degrees. We have become profligate with water as our technology has both enabled and required us to use enormously increased amounts of it. But already there are severe water shortages in many parts of the world, and in North America, including areas with a humid climate, especially in all big cities. Shortages have been greatly worsened by uncontrolled urban, industrial, and agricultural pollution of water sources. The urgency of this problem is hard to exaggerate, but it is past time that water was recognized as a precious and limited resource, and protected and conserved accordingly.

Aside from the oceans, which are not generally considered as water resources because they are not yet economically drinkable or usable for most industrial, agricultural, or domestic purposes, water is present in the ground, as vapor in the atmosphere, and in surface bodies such as rivers or lakes. Atmospheric water may be just as important as ground or surface water; the difference between dry air and moist air can be critical to people and to their livelihood. Atmospheric moisture also has an important bearing on temperature as well as on rainfall and on evaporation rates, which may be as significant as the actual amount of rainfall. Water in the ground is contained or circulating in the soil as moisture, the amount of moisture depending on the climate and on the texture and structure of the soil. Below a certain level under the surface, the ground is usually saturated. This level or line is called the *water table* (see Figure 21), and it follows roughly the contour of the surface, varying from place to place as the soil or rock varies and as the supply of water varies. Ground water from this saturated zone is the source of wells and springs. Where layers of impervious rock retain water, pressure may accumulate on an underground slope as a result of gravity, and a well drilled, or a

WORLD RAINFALL

Centimeters

0 - 50

50 - 100

100 - 200

Over 200

SCALE

GOODE HOMOLOSINE EQUAL-AREA PROJECTION

Map 18 World rainfall

Figure 21 Ground water occurrence

spring naturally occurring, at a low point on this slope may flow or spout without pumping, to form an *artesian well* or spring as illustrated in Figure 21. The saturated zone also helps to maintain moisture in the soil above it, through capillary action, just as coffee rises in a sugar lump when only one end is wet. If the water table is normally low, or if for any reason it substantially falls, as it regularly does to some degree in the drier season of the year, natural vegetation and human activities on the surface may be seriously affected. People themselves, through wells and pumps, may often be the main factor in lowering the water table. If too much water is pumped out of sub-surface aquifers (water-bearing strata), they may collapse or subside (as also in coal-mining areas) and produce surface cave-ins. This has happened in many parts of California and the southwest (see Map 19); in parts of Arizona the water table has been lowered (by surface pumping for irrigation and for urban use) as much as three to four hundred feet.

Water and settlements

With the growth in the last hundred or hundred and fifty years of many cities with a population of a million or more, and with the coming of the modern industrial revolution in large factories consuming enormous quantities of water, ground water has become a less important source than rivers or lakes. Even though it may have to be brought from a great distance, river or lake water provides the more constant and higher

AREAS OF LAND SUBSIDENCE
IN CALIFORNIA   1979

Map 19 Land subsidence in California

volume of supply which large-scale urbanization and industrialization require and which wells alone could not meet. But before the urban and industrial revolution, and in many of the less densely settled or less industrialized parts of the world today, ground water supplies most human water needs except for settlements close to a lake or river. The importance of the water table for agriculture has, of course, remained in any case, but in earlier periods or in the less industrialized parts of the world now, human settlement and all of human activity may be severely limited by what the local ground water can supply. The original sites of many towns and cities were fixed, in whole or in part, by local water. The drier the climate, the more limiting this factor was and is; water is not universally available. In a desert, of course, this limitation is painfully obvious, and there settlement is confined to the few scattered places (oases) where the water table will support wells or springs, or where a river rising in a humid area flows through the desert.

Modern urban-industrial water use

It is not necessary to look only at the desert for evidence of water's influence on settlement, although water's commanding role is especially clear there. In moister climates, too, water is usually an important locating factor for settlement and economic activity, either as ground water from wells, or in the form of rivers or lakes. Pre-industrial people needed much less water but could not move it around easily or cheaply and had to locate settlements and tailor economic activities in accordance with local ground or surface water. Industrial economies require huge amounts of water; they have the means to move it considerable distances, but do not thereby avoid the economic aspect of the problem. Agricultural. industrial, and urban establishments are largely fixed, or at least they can be moved only with great difficulty or expense. The nomad can move when local water is exhausted, but a city, an irrigated farm, or a steel mill cannot; they must pick their sites or adjust to them more selectively.

As individuals, Americans probably use fifty times as much water per capita for their daily needs as did their great-grandfathers. They wash themselves and their property, including cars, more assiduously, mainly because large piped water supplies and household plumbing make it much easier and cheaper. Average Western urban water use in 1980 was about 180 US gallons per capita per day for domestic purposes; the average city of half a million produced in 1980 over 2000 tons of solid waste and about sixty million US gallons of sewage. Daily total water consumption for all uses in the United States in 1980 was well over 2000 US gallons per capita, a staggering fact. But in industrial areas by far the biggest water consumer is manufacturing, and for the United States as a whole, the largest single consumer of water is irrigation. Personal or individual water consumption, although it has increased rapidly, is proportionately small. The figure for industrial use is difficult to make precise, since most industrial operations reuse water more or less continuously, whereas irrigation uses water directly only once. New York City uses over a billion and a half US gallons of water a day. Personal consumption can be somewhat curtailed if necessary in an emergency, but it is much more difficult to reduce industrial consumption unless the industry itself is shut down. Most manufacturing processes require water, and some use enormous amounts. As much as 65,000 gallons, or about 260 tons, of cold, fresh, and relatively soft water is used or reused for cooling and flushing in the making of one ton of steel. Sea water is too corrosive, and some fresh water is too hard or leaves deposits which obstruct pipes. Most of the cooling and flushing water is commonly resused, but in the case of one steel plant near Youngstown, Ohio, which reused water from a small river on which it was located, it was found that the temperature of the river as a whole had been raised to 140° Fahrenheit.

When the growth of population on the west coast of the United States and the strategic and transport problems of the Second World War combined to bring about the construction of a steel plant somewhere west of the Mississippi, Utah was chosen because it was the nearest area to the west coast as a whole which had adequate cheap deposits of coal with iron ore nearby. But most of Utah is a desert, and the most convenient location for the plant, Salt Lake City, could not use the salty water of the lake. The site finally picked, Provo, was a compromise among the factors of coal, iron ore, market and labor, transportation, and water, of which water was the most important. Provo lies near the shore of a large lake, fortunately fresh. But even in humid areas such sites are relatively limited and must be chosen carefully. This is one of the reasons for the location of iron and steel manufacturing along the Great Lakes and on the Ohio and Delaware rivers.

Plants generating electricity from steam-driven turbines, usually the main source of electric power for most big cities, are even more closely tied to water. In total they account for over half of the industrial water use of the United States. Power plants are almost always located directly on a river or lake, from which they take water to cool their condensers. The chemical, rubber, and paper-making industries are also especially large water users. But few industrial operations can afford to locate their plants where ample cheap water is not available. Water is just as essential a raw material to them as anything else they use, and it may be more limiting, since manufacturing needs such large amounts of water that the costs of transporting the water are very high. Industrial demands for water are still increasing rapidly.

Water is also widely used as a cheap and effective means of dispersing industrial and household wastes. But a polluted stream is less valuable or even unusable as a source of water for domestic or industrial use, and pollution may also kill commercially valuable fish and may ruin the stream as a recreation area. Upstream users in this case can effectively destroy the water for downstream users and may even destroy it for competing uses of their own. Alternate means of waste disposal, treatment of wastes before dumping them into streams, or subsequent treatment of the water before reuse are all expensive, but they are clearly essential as the pressure of competing water uses mounts. Many American rivers have been ruined by pollution. Water is far from being unlimited or free for any of the uses to which it is put. It must be husbanded and allocated like any other finite resource whose use involves the danger of exhaustion and requires the consideration of competing demands.

Most modern large cities have outgrown their original water supplies and have had to reach out long distances for water. Los Angeles is often cited as an example of this problem, sending covered aqueducts over two hundred and fifty miles to the Owens and Colorado rivers; Los Angeles also taps water seven hundred miles away from the Feather river in northern California; there are disputes with San Francisco and other potential or would-be users. In any case, transporting water costs money, and it is done only because the Los Angeles area has advantages for settlement, agriculture, or manufacturing which other parts of the southwest lack. As a huge city on the edge of a desert, Los Angeles has a special problem, but it is really no worse than New York's in the humid east. New York must compete for its billion and a half gallons a day with many of the other large eastern cities, since its search for water has largely exhausted what is readily usable in its own tributary area, and New York has begun to tap supplies which Boston and Philadelphia respectively regard as theirs. New York, among many other cities the world over, experiences

IRRIGATED ACREAGE— 1 9 7 8

• = 8,000 Acres Where Irrigation Facilities Existed in 1977
U.S. total = 58 million irrigated acres.

Kilometers
0   500   1000

0   500
Miles

Map 20 Irrigation in the United States

recurrent water crises and may be saved from them eventually only by the development of a process to distill or freshen sea water at a bearable cost. This point has now very nearly been reached, but it seems likely that whatever happens, city water may soon have to be more rigidly metered and sold like any other valuable commodity and at a higher price than is now paid for it.

Fortunately most large population clusters and most manufacturing have grown up in the well-watered parts of the world, and the total economic advantages of these places are also great enough to pay, if necessary at high rates, for the solution of most water problems which arise, although many headaches and maladjustments may have to be overcome and many disputes settled between competing users. In the semi-arid areas of the world, disputes over water rights are chronic; those between competing national states for the use of water from the Jordan River, which flows through Lebanon, Syria, Israel, and Jordan, are an example. On any river upstream use may be detrimental to downstream users. These problems arise, for example, with the Colorado River in the United States, between Arizona, California, and Mexico.

Irrigation

Agriculturally, ground and surface water are important for maintaining soil moisture naturally, or for augmenting it artifically through irrigation. Water for irrigation may be derived from below the water table through wells, or it may be supplied from a greater distance from surface bodies if the agricultural system can pay for it, in the United States, often through government subsidies. Where wells drain too much out of the ground and the water table falls below the limit of the wells, agriculture dependent on wells bumps up against a clear physical limit unless it can supplement its water supply from elsewhere. Lowering the water table may also damage the soil. In parts of semi-arid lowland California where there has been a great expansion of irrigated agriculture based on wells, the water table dropped so far that it was below sea level and allowed an invasion of sea water which poisoned the soils. This was a somewhat unusual situation, but lowering the water table may have equally serious effects on vegetation and crops in many semi-arid areas, where there is the greatest need for wells.

Irrigation from any source pays, even with government subsidies, only if it can support an adequately productive agriculture. Of the total area in western United States which could physically be irrigated by wells or aqueducts only a part actually is irrigated (see Map 20). Only in these places has irrigation so far been profitable, because their soils, topography, and location are favorable enough to make it economically possible for them to pay for the water. They can, and must, grow high-value crops or must get a high return from the land to pay for the irrigation; fruit, fancy vegetables, cotton, sugar beets, or multiple crops of alfalfa as cattle feed all produce a large cash return per acre. Cotton and sugar beets are specially soil-demanding; irrigation for them must be confined to soils which will profitably support them – at least for a time, since accumulated minerals may eventually poison soils in arid areas under "permanent" irrigation – see Chapter 10.

Irrigation is also characteristic of many subsistence agricultural systems, as opposed to the commercial system of North America or most of Europe. In a subsistence economy where there is relatively little sale of the product and most of it is consumed by the producer, irrigation "pays" in terms of the pressure of population on the land and the need to maximize production per acre. Irrigation is confined to areas where pre-industrial techniques can get the water, and in these areas population is very much denser than elsewhere, because irrigated farming can support it. Irrigation may also be applied to areas which are not usually dry but which suffer to a degree, like all of the world, from seasonal and cyclical fluctuations in climate. In well-watered eastern United States, for example, despite its relatively reliable rainfall, irrigation can almost always increase agricultural yields by avoiding the damage caused by dry spells.

Water power

There are two associated aspects of water resources which should be considered here: power and transportation. Hydroelectric power is sometimes spoken of as a "free" resource; this is so only in the sense that it is continually renewed. The world contains a great deal of falling water which might be made to produce power, but only a small fraction of it is so used. The reasons for this are in part physical or technical, but far more they are economic. Potential water power is not evenly distributed over the earth, since it is of course related to climate and to relief; some areas largely lack it, others are well supplied. But where it is physically plentiful it may not be economically usable. It is estimated that about 20 per cent of the total world water power potential lies in one area of the Congo Basin, but almost none of this 20 per cent is used. On the other hand, Japan, with a little over 1 per cent of the total world potential, uses almost all of its 1 per cent and has thousands of times the developed water power of the Congo.

The example of Grand Coulee Dam

An American example may help to explain the difference between Japan and the Congo and differences over the world in the use of water power. There are hundreds of physically possible power sites even in the relatively dry country west of the Mississippi, but only a few of these have been developed. The largest development is Grand Coulee Dam. It is a huge project and cost a great deal of money to build. Why was Grand Coulee chosen in preference to other sites? The source of power is the Columbia River, which derives its water from the relatively humid parts of the Canadian and American Rockies and then flows across semi-arid or desert country, where the dam lies, to the Pacific Ocean west of Portland. Although the Columbia is fed primarily by snow-melt water and is characterized by peak flow in spring and early summer and low volume in winter, its flow is more dependable from season to season and from year to year than is the case with most snow-fed or desert-crossing rivers. The Columbia also carries relatively little silt, since so much of its water is derived from snow or from reasonably well-forested areas; silt can damage turbines and can fill up storage reservoirs behind dams. Storage reservoirs are designed to minimize the effects of fluctuating stream flow, but the problem is easiest and the costs lowest when stream flow fluctuates least. If fluctuation is too great or possibilities for water storage too limited, it may not pay to build a power installation which can operate at capacity only for a few months when stream flow is adequate. At Grand Coulee the river is somewhat confined in a wide gorge, and since it is also losing altitude rapidly it flows with considerable force. Behind the dam site the gorge widens further and easily allows for the backed-up waters which now form a huge reservoir, so that the flow of water through the turbines at the dam can be kept constant. Physically, this is a good place for a power dam.

But power must be sold, and few people live in the arid country around the dam. Power cannot be transmitted economically in most cases for more than about three hundred miles. Beyond this distance too much power is lost from the lines, or too many booster stations are necessary to make transmission pay. Power also cannot be stored economically in any significant amounts. Since there are high fixed charges on the dam in the form of interest on the huge investment and of maintenance, the plant must produce and sell power continuously. It cannot afford to shut down or to run at a fraction of capacity. Hydroelectric establishments, in other words, are economically possible only if they have within a range of roughly three hundred miles a large market with a high and constant demand for power, or what is called a high constant load factor.

Well within a 300-mile radius of Grand Coulee lie nearly all of the large cities and manufacturing establishments of the far west, north of California. The urban and industrial area of Seattle-Tacoma is about 160 miles away, Portland about 250, greater Vancouver (as large as Seattle) about 210, and Spokane about 75. Because Grand Coulee can deliver large amounts of power to them, many new, heavy power-using industries have been attracted to these cities and thus have helped to keep the costs of power down. The larger and more constant the sales of power, the lower the unit costs (the cost of one unit of power). Population growth in the cities served by Grand Coulee, based largely on new or expanded industry, has thus had a tendency to snowball. Industries which require large amounts of power (for example, aluminum refining and the smelting of certain ores) have become prominent in these cities since Grand Coulee power has been available and are now an areal specialty. As pointed out in Chapter 2, one person employed in a basic industry is usually accompanied by about two people employed in the service industries which bake the bread, clean the streets, staff the professions, operate department stores, or supply materials to the basic industry. Counting family dependents, it is estimated that one worker in a basic industry usually results in about eight other people dependent on him, serving him, or dependent on those who serve him and his dependents. In this way Grand Coulee has contributed substantially to the growth of population in the Pacific Northwest. This may be attributed in part to water resources, but weight must also be given to the force of location or access and to the existing or potential markets which the dam could reach. Little manufacturing has gone to the dam itself. Manufacturing has settled in established urban centers which were there before and which are based on a variety of other advantages, including labor force, transportation, access to productive hinterlands, and favorable site.

Most of Grand Coulee's revenue comes from power sales, but this was not enough to meet the costs of the dam, especially at the time it was planned and built in the 1930s, when future power sales, the impact of the Second World War, and the growth of population were greatly underestimated. Some of the water is also used for irrigation, and sales of irrigation water were expected to help meet the cost of the dam. Irrigation of this part of the Columbia plateau can be justified in part because the soils are exceptionally good. The parent rock is volcanic lava, and the semi-arid climate has allowed a fertile and well-structured soil to develop. Although water from the Columbia has to be pumped up out of the gorge about three hundred feet to the level of the surrounding plateau, irrigated agriculture on the plateau is very productive. The irrigated farms concentrate on high-value crops, mostly multiple crops of alfalfa and vegetables, and they realize a high return per acre. Thus soil, as well as location and site, helps to explain why a dam at Grand Coulee was possible.

Most modern dams are multi-purpose projects whose costs are met through a variety of uses. Power and irrigation are usually the most important, but flood control, the maintenance or improvement of navigation, water for household or industrial use, and recreation on the artificial lake which may result behind the dam are also common benefits. Usually it takes a combination of uses to pay for, or to justify, the building of a big dam. Some of them, like recreation or flood control, are hard to put a money value on, but they must be considered. The general effect on the economy of the area of all these things taken together is also worth something; it will increase the population, raise the average income level, and make the area a pleasanter or more profitable place to live. These indirect benefits should be calculated and some of the dam's costs charged against them. The extent to which all or any of these products of a dam can profitably be used (and thus whether the dam can be built) will depend on where the dam is and what kind of country and people lie around it or have reasonable access to it. Water resources in this sense clearly involve much more than the physical fact of falling water.

Water transport

Water transport is usually the cheapest way to move bulk goods. Friction is slight, and costs of building or maintaining the right-of-way are low compared to most forms of land transport. But most important is the fact that the carriers, or their capacities, can be very much larger than on land without being proportionately expensive to operate. Generally, the larger the carrier, the cheaper its unit haulage costs, assuming the carrier can consistently make up a full load. There is a point of diminishing returns beyond which increasing the size of the carrier does not bring lower unit costs, but the *economies of scale* (the bigger, the cheaper) work in favor of water transport. If carriers, by land or water, can be built to facilitate mechanical loading and unloading or to haul only one type of cargo which can make maximum use of the carrier's capacity, costs are further reduced. The largest single cost item of transport, especially for large shipments, is often loading and unloading, known as *terminal costs*, rather than the actual movement along the line, known as *line costs*. Water carriers often achieve economies in this respect, particularly in specialized ships like tankers, or ore, coal, or grain boats. General cargo is more expensive to load and unload and to carry, but even for general cargo most water carriers have terminal cost and line cost advantages over most land carriers. In recent years most ocean-going ships have been designed as container carriers; huge containers holding 20–60 tons of cargo (roughly comparable to freight cars) are stacked into and on the ship and are loaded and unloaded mechanically, for transfer without unpacking to and from railroad cars or large trucks. Another important factor is length of haul. Since terminal costs may often be greater than line costs, long hauls are often disproportionately cheap, if a full load is available for the entire distance. Long hauls usually cost less per mile than short hauls, especially for large carriers. Most ocean hauls are necessarily long, and ton-mile rates (the rate for carrying one ton one mile) can therefore be very low.

Waterways and economic development

Before the industrial revolution and the building of railways, water transport was the only economically practical carrier for bulk or low unit value goods moving beyond a few miles. Transport by cart or pack animal may within fifty miles double the cost of the goods carried unless they are of very high value per unit. In pre-industrial periods or areas, settlement and economic development are particularly tied to waterways. Rivers were the almost universal means of exploration and settlement in colonial areas: witness the roles, to give only a few examples, of the Hudson-Mohawk, Ohio, Mississippi, Connecticut, James, Columbia, and St Lawrence in North American history, the Amazon, Congo, Nile and Volga for their areas, and in the East the Ganges, the Irrawaddy, and the Yangtze, on which Western economic colonialism so massively rested. With the coming of the modern industrial revolution there was a tremendous increase in the volume of exchange and a great need for cheap carriers to move huge amounts of low unit value goods like coal, iron ore, or foodstuffs. Water transport has increased greatly in absolute terms where industrialization has spread, even though with the building of railways water transport has declined relatively. But the Rhine, for example, still carries a greater volume of traffic than the combined railways which parallel the river, although German railways as a whole haul more goods than internal waterways.

Some smaller areas which originally were nourished economically by waterways, Cape Cod, tidewater Virginia, or parts of the Ohio valley, for example, have declined when railways or roads short-circuited traffic around them. There are ghost towns and rotting wharves in many such areas, where a few miles away the railroad and the automobile have stimulated new towns. Most cities founded before 1860 necessarily grew up around or along the water routes which served them. These older parts of many cities have now become relatively less prominent and may even be decaying as transport and economic activity have increasingly centered on railroad termini or motor routes. But frequently the railway termini, at least for freight, have grown up in close association with the original waterfront, where cheap access and break-in-bulk points are concentrated. Passengers can now travel more quickly than by water, and their termini are often separated from the old waterfront where the entire life of the city originally centered. Such local or small-scale readjustments

Figure 22 Hypothetical settlements along a river

should not obscure the continuing importance of water transport, however. In the three greatest centers of manufacturing, Japan, Europe
(including the Ukraine), and eastern North America, naturally available
water routes have been a strong advantage, an important reason for early
industrial leadership, and a continuing stimulus to economic growth.

Settlement on rivers

Settlement and economic activity are seldom uniformly spread along a
river. The effects of the fall line have been discussed in Chapter 5, and
similar cases can be found in many areas outside the eastern United

States. There are usually a variety of other break-in-bulk points along a river where carriers change (see Figure 22 on page 159). At or near the mouth is perhaps the most common, since most rivers are not navigable for ocean-going ships much above their mouths. The most economic ocean carriers tend to be the largest, and it usually does not pay to haul goods all the way across the ocean in a boat small enough to use even a big river. Most rivers are settled particularly densely near their mouths, where level land and fertile alluvium are also likely to be concentrated. An atlas map of the United States or of the world will show how close a relationship exists between large cities and river mouths. In Europe the mouth of the Rhine supports three large cities (Rotterdam, Amsterdam, and Antwerp), as well as most of the Dutch economy, though not entirely on the basis of Rhine traffic. But there is usually a succession of heads of navigation farther upstream: one for ocean-going vessels, one for large river boats, one for smaller river boats, and so on. At each well-marked break, which may be a falls, or rapids, or a narrowing of the river or its valley, a town or city may arise, taking advantage of the opportunities which any break-in-bulk point provides. Such a city has to consider access to the land also, and it is therefore unlikely to be in a gorge, for example, but downstream from the gorge where there is a wider productive hinterland and easy access routes to it. For example, Montreal is larger than Cornwall, Ontario, where the gorge and rapids of the St Lawrence begin, and Montreal is also straddled by the mouths of the Ottawa River where it joins the St Lawrence.

Confluences are favorite settlement sites, as Montreal suggests. The main stream is likely to be less navigable above the confluence than below, and there may also be a change of carriers between the larger stream and the smaller one. There will certainly be a focus of routes and opportunities for access in different directions, and the rivers may have helped to create a broad productive lowland. Philadelphia, where the Schuylkill flows into the Delaware, is an example; Pittsburgh, where the Allegheny and Monongahela rivers join to form the Ohio, is another. Cincinnati, St Louis, Omaha, and Minneapolis-St Paul are also confluence cities. As with many cities at river mouths, however, the exact point of the confluence may be avoided; often it may be subject to floods, and higher ground may be available near enough to take advantage of the confluence.

At the river's mouth depth may be adequate for ocean-going ships for a space upstream, and it is cheaper for ships to go up as far as they can as long as they do not run into expensive delays. Silting and sand bars may be less of a problem upstream or on a downstream tributary of the main stream, as at Shanghai or Calcutta, than at the edge of the delta. The lower delta may also be more subject to floods or may have permanently sodden or salty soil. New Orleans is about a hundred miles from the mouth of the Mississippi, although there is no question of a fall line near New Orleans, nor near Hamburg, which is over fifty miles from the mouth of the Elbe.

Many cities so placed have had problems, because the size of ocean-going ships has increased particularly rapidly since the 1890s. Some cities have established *outports* at the new downstream head of ocean navigation where the larger ships load and unload. Cuxhaven is the outport for Hamburg, Bremerhaven for Bremen (haven is a variation of *hafen*, the German word for harbor or port), Le Havre for Paris (Rouen), and Gravesend-Tilbury for London.

London illustrates another point on a river where settlement tends to cluster: the site of a bridge or ford. Often the lowest point on a river which can be bridged will also be the head of ocean navigation, as it is with London, with several of the cities of the fall line in eastern United States, and with many other urban centers elsewhere. But any bridge site is likely to be a good break-in-bulk point and assembly-distribution center, since land routes will be bunched together there in order to get across the river. "Bridge" or "ford" in the name of a city may reveal one of the city's site advantages. It does, for example, with Hartford, Connecticut, Oxford and Cambridge in England, Klagenfurt and Innsbruck in Austria, Saarbrücken in the Saar, and Bridgeport, Alabama. Familiarity with maps and the study of place names can be fascinating.

12 LAND FORMS AND SETTLEMENT

This last of the five interrelated elements of the physical environment is perhaps the only one which can rival climate as an influence on people. Land forms may arbitrarily but conveniently be divided into four categories: plains, hills, plateaus, and mountains. Hills are not only lower than mountains but are usually more rounded and more gently sloping; plateaus are simply elevated plains. Elevation, however, is less important than *local relief*. Local relief includes the degree the slope of the ground, how much difference there is in feet between the highest and the lowest parts of an area, and how much of the area as a whole is level or sloping. A highly elevated plateau may be more usable than a low-lying area with heavy local relief. Many areas labelled mountains may actually contain extensive sections of low relief, and many plains may be interrupted by broken country.

Mapping relief

Most relief maps of large areas rely on altitude tints to show elevation above sea level, often using green for elevations under a thousand feet and darkening shades of brown, purple, or red for the higher elevations. Altitude tints may obscure great variations in local relief. They may also, as in the case of the North American Great Plains, suggest topographic barriers where none in fact exist. The map maker must draw lines between altitude tints at some more or less arbitrary elevations. If a large area with widely varying elevations must be mapped, each color category must cover hundreds or even thousands of feet within which there may be considerable local relief which the altitude tints cannot show. A map of India, for example, must include both the Himalayas and the Ganges valley, but at normal page size such a map would be difficult to read if it used more than eight or ten categories of altitude tints; each color would thus have to represent one to three thousand feet, and each would therefore conceal local variations which might be only a few feet less.

More accurate relief maps of smaller areas are made using multiple *contour lines*, as in Map 4 in Chapter 3 on page 37. (A *contour line* connects points of equal elevation.) The smaller the *contour interval*, or the difference in elevation between two contour lines, the more accurate the map, and the more difficult it is to read. The smaller the variations in local relief, the smaller the contour interval which can be used and still leave the map reasonably clear. Where small local variations in relief are especially important, such as in irrigated areas or the centers of cities, a contour interval of five feet or less may be used. For larger areas or areas with heavy local relief, an interval of fifty or a hundred feet or more is appropriate. What is called a *vertical profile* may also be constructed on the basis of a contour map, as illustrated on Map 4. For a large area small-scale contour maps are almost illegible or unworkable unless generalized so much that altitude tints might just as well be used instead. The best small-scale relief maps of large areas make use of a combination of altitude tints and what is called *plastic shading*, now used in most atlas maps of land forms. This visually suggests the land forms without losing any accuracy and makes the map much more vivid and less flat looking. But no small-scale relief map can be accurate in detail, and it can give only a highly generalized picture.

Relief and population

Nevertheless, a comparison of the world physical relief map (page 164) with the population map (page 89) will demonstrate a clear association between people and plains. Only in the tropics or in South Africa do large numbers of people live in the highlands. This is not so much because in the tropics climate is pleasanter in the highlands, but because the highland climate offers greater economic opportunity. Desert, arctic, and equatorial plains are often avoided; climate overrules land forms. Otherwise there is remarkable consistency. Japan, which the world map makes seem an exception, actually is not; the Japanese highlands are sparsely settled and the lowlands almost uniquely crowded, although most world maps are too small to show this adequately. The combination of altitude tints and small scale plays tricks in a few other places, most notably in southeastern Spain and in the Chinese province of Szechuan; both actually contain extensive areas of low relief. Notice on the population map how the Alps, the Pyrenees, the Rockies, the Himalayas and their many associated ranges, the Caucasus, and many lesser mountains stand out as empty areas. Why is there this general avoidance of mountains and concentration on plains?

Lowlands and mobility

Lowlands have a number of obvious but vital advantages. Circulation and

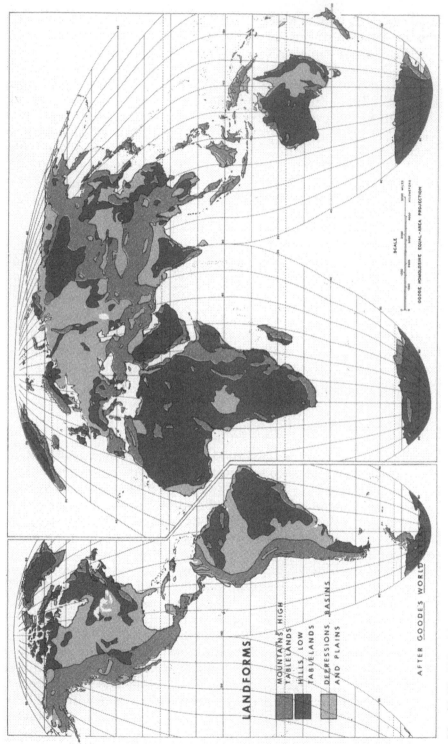

LANDFORMS

MOUNTAINS HIGH
TABLELANDS

HILLS LOW
TABLELANDS

DEPRESSIONS BASINS
AND PLAINS

AFTER GOODE'S WORLD ATLAS

SCALE

GOODE HOMOLOSINE EQUAL-AREA PROJECTION

Map 21 World land form types

mobility, conditions of economic development and of resource use to support large populations, are maximized on plains. Mobility means that areal specialization can proceed – that each place can maximize its particular advantages and at the same time receive the special products resulting from the advantages of other places. If everyone had to grow his own grain, bake his own bread, and make his own alarm clocks, we would not get very far. Many mountain people do have to do roughly this, but on the plains cheap circulation helps to make it unnecessary. The division of labor (job specialization), dependent upon exchange, is a universal basis of economic growth. Manufacturing, especially heavy manufacturing, is absolutely dependent on cheap exchange. Its bulky raw materials cannot be moved around otherwise, nor can its finished products be distributed. Where manufacturing is established on the plains, dense population is likely to result, as basic labor and to provide services. This further increases the market and attracts commercial production of many kinds, including commercial agriculture in the market area. Mobility over the plains is not due only to the absence of topographic barriers. Rivers flowing across relatively level areas are more likely to be navigable than mountain or hill streams; the significance of cheap water transport has already been discussed. In many cases the rivers have helped to create the plains by eroding away the original mountains or by depositing silt to fill up ancient valleys or coastal seas. Actually, many of the great plains areas of the world either fringe the sea, where rivers have extended the land by deposition, or lie open to the sea. Coastal plains have attracted the largest blocks of dense population, as shown on the population map, where the circulation advantages of the sea are available as well as the mobility of the lowlands. A great deal of the world's heavy manufacturing is concentrated on coastal plains.

Plains agriculture

Manufacturing (including the services attached to it), and exchange or commerce are two of the three major economic activities requiring and contributing to dense population. The other is agriculture. The remaining activities – herding, hunting, forestry, fishing, and mining – are either extensive in nature, are concentrated in marginal or severe climates, or may have no close connection with major settled areas. Lowlands offer overwhelming advantages for agriculture. The many problems of cultivation on slopes are avoided. The soils of the plains are nearly always superior and may also have benefited from alluvial deposition. Outside the tropics and if the sea is near, as in many plains it is, climate is likely to be moderate, in both rainfall and temperature, compared to that in the highlands. In the two most heavily peopled lowlands in the world, in India and China, these agricultural advantages are the primary attraction and basis for settlement. But even in industrial Europe and North America the agricultural virtues of the plains are of great importance. Commercial agriculture depends on market access just as manufacturing does, and access and markets are maximized on the plains, where most people live. A dense population requires a productive agricultural system, whether its nature is commercial or subsistence. The availability of surplus food and the presence of the relatively dense population associated with productive farming areas are further attractions for most manufacturing, which must obtain and feed its labor and must have access to consumers as well as raw materials. As the eighteenth-century historian Edward Gibbon put it, in a perceptive appreciation of the geographic concept of spatial relations, "The first and most natural root of a great city is the labor and populousness of the adjacent country, which supplies the materials and subsistence of manufactures and of foreign trade."[1] Populousness and productivity from agriculture or manufacturing, plus easy mobility, mean that lowland areas are also likely to be generating centers for trade and foci of trade routes.

Mountain settlement

One may ask why anybody bothers to live in the highlands. Most people like their native places, and often the more different from adjacent places the more passionate the attachment. If one's native place is economically poor, its other virtues are certain to be extolled, or poverty itself may be made a virtue. Mountains are physically distinctive places especially sharply differentiated from lowlands, and those who live in the mountains often assume that they have the better of the bargain, even if not for economic reasons. In practice, mountains make good refuge areas, and many mountain people originally came there because they were pushed off the better lands by stronger or larger groups. As with the rain forest, security is the result not only of physical barriers but of the physical and economic poverty of the mountains, which attract few contestants. Religious, social, and political dissidents may also swell the mountain population, and there are always a few hardy souls who honestly prefer the highlands as a place to live. In many mountain areas human settlement is too old, or too little is known of its origins, to say for certain whence came its inhabitants or why they went there. But with few exceptions outside the tropics, they are not there because the mountains offered them greater economic opportunity than the plains.

Poverty may also be reflected in emigration. Most mountain areas send out a constant stream of people seeking better opportunities elsewhere. Banditry or organized raiding of more favored places may also be a highland specialty, in Europe in the past, in some other parts of the world still. Mountain conditions – clan rivalries or the need for defense against pressure from the lowlands – may encourage the development of military prowess, and economic hardship may provide an incentive for the use of this prowess. Mountain and plains peoples seldom get along well together, and this further tempts the highlanders. The hills are frequently sources of trouble, either for the lowland trading town or farmer who may be raided or for the government anxious to preserve order. The shoe may be on the other foot when the economic and political power of the plains reaches out to subdue or control the mountains. History is full of such two-way struggles, and where highlands and lowlands meet, the landscape may be marked with fortresses and battle sites, such as Stirling in Scotland and Carlisle in north England or Carcassonne in southern France. The spate of movies, short stories, and novels about military adventures in the North-West Frontier Province of India or on the borders of Scotland are good melodramatic reminders of such struggles. The plains people are far from guiltless in this and have often regarded the highlands as fair game.

Mountain economies

Mountain cultures and economies are as distinctive as their physical base. Heavy manufacturing and field agriculture, except in chance pockets of level or fertile land, are either excluded or severely limited. This may leave scope for dairying or for the raising of sheep and goats for wool or skins. Sheep and goats will tolerate slopes and relatively poor pasturage better than cattle will. Their meat may be consumed locally but can seldom be exported; it is too low in value per unit of weight and too perishable to be carried profitably by the transport available. The wool is often made into cloth before export for the same reason, and transport costs also explain the making of cheese. Handmade manufactures are prominent, as in Switzerland, due to a long slack season and to limited local resources in both a physical and economic sense. Value added by manufacture is stressed, and hand-woven woolens, for example, may command a high price, although they seldom return a very high hourly wage or profit to the maker. Bulky or perishable exports are difficult. Grains tolerant of the poor soils or harsh climates common in mountain areas may be grown on more level patches of ground for local subsistence or for animal fodder. To be exported, grains must usually be converted into a higher-value form as animals, for example, or as whiskey; Scotland and Kentucky still illustrate this point. In very high mountains specially adapted animals may be used: llamas or alpacas in the Andes, yaks in the Himalayas. All of these animals are physiologically suited to high altitudes, and indeed do not thrive below about ten thousand feet. In Tibet and the high Andes they are the preponderant source of food, clothing, transport, fuel (in the form of dung), and exports. One of Tibet's exports for many years, in addition to wool, was yak tails, high in value by weight and virtually imperishable, to be used as fly whisks in China and India and as false beards and Santa Claus whiskers in the United States and Europe. This is a good measure of mountainous Tibet's poverty and of its basically subsistence economy, in common with many mountain areas.

Tree crops

Slopes need something which will hold the soil, and bush or tree crops may be as good as grass or may thrive as well. Two particular tree or bush crops grown almost exclusively in the hills are tea and coffee. Both tolerate slopes, need good drainage, and profit from the coolness and moisture of higher elevations. With tea this is more a matter of economics: the lowlands are taken up by the more profitable and more particular crops. Where it is grown on hill plantations, as in India or Sri Lanka (Ceylon), tea may support a good many people locally. In China and parts of Japan it is a largely untended, casual crop, visited by lowland people only for occasional pickings. Coffee is grown almost entirely in the tropics, since the tree and its fruits are damaged or killed by frost. The coffee tree suffers in excessive heat, however, and so does better in the highlands, where slopes do not especially bother it. Tea and coffee are probably the most important commercial crops of the mountains over the world as a whole. If the sparsely settled mountains are forested and the adjacent densely settled lowlands denuded, as is often the case, lumbering may be another source of livelihood, but the transport problem is particularly pressing. Areas where the economic distance to market is great must concentrate for export on high unit value goods. Timber is not high in value by weight, and the ruggedness and remoteness of mountain areas often mean that timber there is not exportable, unless it can be made into something locally which can bear the heavy transport costs. Wooden Swiss cuckoo clocks can, and so can Panama hats, which despite their name are made (the best of them) in mountainous, remote Ecuador, where there is need for an export whose high unit value comes mainly from human labor, since there are few other local resources.

Water power and minerals

Potential water power and mineral deposits are commonly found in mountains, but only under certain circumstances do they contribute much to the density of population or to the level of the local economy. In many cases they may not be usable at all. Water for uses other than power may often be derived from mountains for the supply of a less humid nearby lowland, as in the Mediterranean Basin as a whole, or the Wasatch piedmont oases of Utah. But the water is used in the lowlands, not in the mountains, important as the mountains are in this respect for the support of the plains people. The conditions under which water power becomes a usable resource have already been considered in the preceding chapter, and it should be clear that while falling water may be localized in the mountains, this means that it will often be beyond economic reach of an adequate market. In Japan it is not; plains and mountains lie close together, and overall distances are relatively small. A huge nearby industrial market provides a high constant load factor. In Switzerland hydro power can be developed to support the diversified Swiss industries. In France, Germany, Italy, and Scotland, mountain sources of power are near enough to large markets. But in many mountain areas this is not the case; even where it is, the power is likely to be shipped out to where the people are rather than being used in the mountains, and this is in fact the almost universal pattern. Switzerland in this as in other respects is an exception to the rule.

It is much the same with minerals, which are often associated with or exposed by the folding and faulting of the earth's crust, the basis of most mountains. Especially if they are minerals low in value per unit of weight, like coal, they may be too remote, measured in economic distance to be usable. If they can be brought out at a bearable cost, they will be brought *out*, not turned into manufactured goods in the mountains where they occur. The lowlands have a virtual monopoly of the other advantages for manufacturing, and they draw minerals to them.

Tourism

Mountain areas are often attractive as resorts for people from the crowded plains, especially where the plains are hot in summer. But situation is usually more important in accounting for the presence of tourists than the grandeur of the scenery or the cool weather. The Poconos and the Catskills, hardly in the mountain class, attract annually millions of vacationers from nearby New York and Philadelphia, while the infinitely more impressive ranges in the Himalayas or Alaska rarely see visitors. Nearly every even slightly elevated or hilly area near a densely populated urbanized lowland derives a large and increasing share of its income from tourism, especially where transport facilities are well developed and average lowland incomes are high enough to permit luxuries such as vacations.

Mountains and transport routes

The cost of transportation is a crucial factor in mountain economies. Transport routes generally avoid mountains, not so much because of physical barriers but because the mountains are relatively unproductive. Sparse settlement on a low economic level cannot support a railroad and may not even be able to support a road. It is true that land transportation, especially the bulk carriers like railroads, must be sensitive to grades. But where there is an economic basis for exchange, land routes and even railroads can afford to penetrate mountains despite the physical barriers and consequently increased costs, as for example in the coal areas of the Appalachians or through Switzerland. For the most part, however, transport routes penetrate mountains primarily to get to the other side. When roads or railways cross mountains, they try to combine a physically easy route with a commercially productive one. This is the basis of mountain passes. Some passes have been chosen principally or solely for physical reasons, but in most cases economic considerations are also involved, if not in the existence of the pass, then in the degree to which it is used or in

the choice of alternate routes, depending on what productive or less productive lowland areas are connected by the routes. Settlement is not
likely to concentrate in the pass itself; there will be few functions for a
settlement to perform aside from sweeping the snow or landslides off the
road or providing St Bernard dogs for benighted travelers. A better place
for settlement is at the lowland approaches to the pass, where there may
be a need for a break-in-bulk point, for a strategic or military post, or
simply for an assembly and distribution center where lowland routes are
gathered into a knot before ascending the pass. Good examples of such
settlements would include Milan, Denver, and Cheyenne. Trade is
generated and handled on the lowlands.

Generally, transport routes penetrate mountains in order to get across
them, because the mountains separate complementary areas between
which trade flows. Occasionally roads or railways will do more than get
across the mountains and may spread their lines out to haul minerals
down to the lowlands. There was a time, for example, before about 1930,
when there was a greater development of railroads in the mineralized
areas of the Rocky Mountains of the United States than on the surrounding arid lowlands, which by comparison offered fewer cargoes. But this
simply emphasizes the point: transport routes go where there is something to haul, whether it is in the mountains or not. Most mountains have
little to haul, and this contributes to the high cost of hauling that little,
whether it has to go on a porter's or an animal's back, or on a one-train-a-
week narrow-gauge railroad.

Transport flow and transport costs

A high and constant volume of flow helps to keep unit transport costs low,
just as greater capacity of the carrier makes for lower unit costs. The
larger the capital investment in right-of-way, carriers, and capacity, the
greater the need to use these facilities to capacity in order to pay off the
high fixed costs. A large flow is necessary to justify the initial cost of building a road or railway and to maintain it in profitable operation. It may cost
several million dollars a mile to build a first-class railroad or a modern
interstate. Where a large and constant volume of traffic makes it possible
to provide such facilities, ton-mile costs for railroad haulage, for
example, may be only 1 or 2 per cent of the costs of human or animal
porterage in areas where there is not enough present or potential traffic to
support a railroad. Human porters and pack animals, which are common
in many mountain areas, are the most expensive of all carriers because
their capacities are so small. Mechanized carriers of large volume, such as
railroads, benefit from the economies of scale. Where the flow is mainly a
single bulk commodity, it may be possible, as with ships, to use cars
designed especially for that commodity, which can be loaded and
unloaded mechanically, and ton-mile costs will fall even lower. A train
may haul 5000 tons of general freight at $0.05 per ton-mile or 8000 tons of
coal at $0.03 per ton-mile, whereas a pack animal may carry 200 pounds at

$0.80 per ton-mile and a human porter 100 pounds at $1.50 per ton-mile. The economic range of porters is also severely limited. If the volume of flow is small or irregular on a road or railroad, the facilities and total capacity will have to be limited if fixed costs of maintenance are not to swallow up all the profits; the average unit cost of haulage will consequently be higher than where large and constant volume makes larger facilities and capacities possible. Where the volume is too low to justify even road building, exchange will have to depend on still smaller carriers, cheaper to build or to operate per mile but much more expensive in ton-mile costs.

Transport costs may also be affected by the predominant direction of flow. Where most of the weight and volume is moving in a certain direction (the "heavy" direction), it may be possible to offer lower rates as well as more frequent service in larger amounts in the heavy direction. A carrier may often offer lower rates for freight moving in the "light" direction, however, in order to attract business and to meet fixed costs, especially if the carrier must run in the light direction anyhow in order to pick up freight which it will then move in the heavy direction. It is expensive to operate a carrier empty, and *return-haul* arrangements and special rates often attempt to compensate in such situations. Mountains usually send out a greater weight and volume of raw materials than they import as manufactured goods, and may therefore pose return-haul problems for carriers, sometimes reflected in higher rates. Similar relationships elsewhere, between agricultural or raw-material-producing areas and manufacturing areas, as between western and eastern United States as a whole, may affect transport rates. Finally, mountains pose obvious physical problems for any carrier, which are also reflected in transport costs. For all the above reasons, mountain transport tends to be expensive and thus further limits economic opportunities in the mountains.

Mountain cultures

Mountain cultures can perhaps be summed up in the word "backwaters," but this would be unjust to many areas, or at least to many aspects of them. Switzerland is obviously the reverse of a backwater, despite the persistence of the Romansch language. But Switzerland is virtually alone among highlands in the degree of circulation and outside stimuli which it enjoys. Generally, these are at a minimum in the mountains; while that may be eminently desirable for some tastes, it undeniably tends to preserve archaisms. In speech, dress, customs, attitudes, and every other cultural attribute, most mountain people are usually out of gear temporally with the lowlands. This may often be a blessing, especially from the mountain point of view. But although isolation may foster local feeling or keep worthwhile traditions alive, it also tends to mean that change is slow in the mountains and that they lack the dynamic quality of the lowlands, where circulation and external stimuli are concentrated.

In the mountains of Kentucky and Tennessee some people still sing

Elizabethan folk songs long since forgotten in England, and their speech
is said to be much closer to the speech of Shakespeare's England than is
the contemporary Londoner's. The mountain fastness of the Pyrenees
long preserved the Basques, whose origins have been obscured by time,
but whose language and culture are unrelated to any in Europe. They
once occupied a larger area, including adjacent lowlands, and the mountains probably became for them a refuge area, where they now support
themselves mainly by sheep raising. The tiny state of Andorra, also in the
Pyrenees, owes its continued independence largely to the mountains. In
Tibet Buddhism (and a variety of archaisms) was long preserved,
although Buddhism has all but vanished in its Indian home and in China,
where it once flourished. Mountainous Ethiopia, early Christianized by
missionaries from Egypt, kept its religion, though surrounded for
hundreds of miles by other creeds, and today it may be closer to some of
the forms of fourth- or fifth-century Christianity than is Rome itself,
where much has happened since. Ethiopia's political independence
throughout the age of colonialism in Africa (except for the brief Italian
interlude) and despite several attempted conquests is also attributable
in large degree to the physical protection and undesirability of the
mountains.

Distant effects of mountains

Aside from human use of or adjustment to highlands, mountains may also
exert their effects upon other areas, sometimes at a considerable
distance. In Chapter 8 some of the climatic effects of mountains in
creating rain shadows on one side and heavy rainfall on the other have
already been discussed. Mountains usually have a strong influence on
temperature also, shutting out maritime or continental influences. The
Alps, for instance, plus the adjacent Massif Central in France, help to
account for the warm sunny winter climate of the Riviera; the Himalayas
shield India from the cold outblowing winter winds of Inner Asia and help
to give India a virtually unbroken growing season; the bitter winters of
lowland Montana and Idaho or of eastern Washington may be blamed in
part on the Cascades and the Rockies, which shut out the moderating
influence of the sea.

Any area surrounded by mountains, or with one of its approaches
blocked by them is likely to be profoundly affected as a result, quite apart
from the climatic effects. Much of the distinctiveness of Spanish civilization is due to the barrier imposed by the Pyrenees, which are particularly
effective because there are few easy routes across or around them, unlike
the Alps. Spain is mostly mountains, at best a plateau, and the Pyrenees
merely reinforce general cultural and economic isolation, hence the old
expression "Africa begins at the Pyrenees." Within Spain there is still no
common language on the scale of most national units, and there are very
strong local regionalisms; to be a Catalonian or a Castilian in speech and

in origin is often more important than to be Spaniard. Political separatism, archaic culture, and limited economic development in the Balkans may also be attributed in part to mountain barriers.

Topography (relief) may also help to deflect or to channel movements of ideas and of goods. Settlement and transport follow the lowlands, substantially penetrating the highlands in most cases only to get to the other side and making use of the lowest or most convenient routes. Mountain barriers may thus focus trade and affect a very large area. New York City, as already analyzed, owes much of its size and importance to the Mohawk corridor through the Appalachians, which focuses trade between the Middle West and the Atlantic coast. On the Pacific coast of the United States, Seattle, Portland, San Francisco, and Los Angeles are all close to easy passes or river valleys leading east across the mountains. These sites are the only ones west of the Rockies which support large cities.

Local relief and urban sites

On a smaller scale than has so far been discussed, slope and elevation are usually of local importance as one aspect of site. A city or town, for example, may choose an elevated site to avoid floods, to moderate climate, or to gain military protection. Defensibility was a major concern for nearly all settlements until relatively recently, since only a few favored areas were free from chronic civil disorder or invasion. Only within the last two or three centuries have settlements been able largely to ignore topographic protection, or to omit walls or moats if the city had to be sited on level ground. Although elevated sites continue to have some advantages in flood protection and for residential use, trade and manufacturing suffer on such sites, and most large modern cities avoid high or rugged ground, at least for their business centers, if level ground is available. Many cities on elevated sites have retained their old centers, often built around a cathedral or castle on the fortified hill, and have developed a new commercial, manufacturing, and transport center on the adjacent lowland. Where enough level land is not available, or where it is less favorably located (Seattle, San Francisco, Boston, Pittsburgh), land may be reclaimed from the sea or river to support manufacturing and transport, especially for large space users like railroads, which often arrived on the scene after most of the existing level land near the center of the city was already occupied. (See Maps 6 and 22 showing made land in Boston and Seattle.)

Residential housing can be more tolerant of slopes and may even seek them, particularly high-rent apartments and high-value single-family houses. In most American cities the higher or more rugged the land in the suburban area, the greater its price per square foot and the more expensive its housing. Such land is away from the commercial and industrial nuisances of the city's center, which depend on access and therefore seek level ground. Higher ground, if it presents problems for bulk transport,

Map 22 Made land and land use in Seattle

may also have escaped deforestation and may still be wooded; it is certain to have a pleasanter micro-climate than the lowland center of the city, especially in summer. If slope makes the area less accessible, so much the better for those who like privacy and are willing to pay for it.

Agriculturally, slopes may be used for fruit or nut trees or grapes, which require good drainage, do not suffer on slopes, and may be able to maximize sunshine on hillsides facing south. The famous Rhine wines come from grapes grown on such hillsides in the cool climate of the Rhine valley, where the slopes facing north are usually uncultivated. Similar use is made of hillsides in most mountain areas. But such local advantages or uses of slopes should not draw attention from the larger fact that in general, highlands and poverty go together. People are predominantly creatures of the plains.

NOTES

1. Edward Gibbon, *The History of the Decline and Fall of the Roman Empire* (London, 1787), chapter 71.

13 RESOURCES AND POPULATION GROWTH

Resources defined

The word "resources" is often taken to mean agricultural and industrial raw materials. It is sometimes implied that soils, water, mineral ores, power, and fuels form the sum of all resources, and also that every lump of coal or drop of oil in the world adds something to the total resource inventory. There is no more reason in this than in saying that food means the seeds of plants and that all plant seeds, including the poisonous or inedible ones, are food; or that literature means poetry and that all doggerel is literature. Confusion is less likely if the word *resources* is given the meaning which general usage and dictionaries give to it: things on which people depend or which support them. Resources have meaning only in reference to people.

But mankind is always changing, and human changes are accompanied by changes in what constitutes a resource or in what people depend on. This is different at different periods and in different places at the same period. Each civilization rests on a different complex of resources and attaches differing importance to the things which it uses. Modern industrial civilization depends most importantly on power – from mineral fuels, from water, and from fissionable elements. The development of these new sources of power distinguishes the modern industrial revolution from the periods which preceded it, when minerals and many other things on which people still depend were being used.

The basic resource of all pastoral peoples, and even of commercial stock raisers to an extent, is grasslands. Exploited through herds of animals, grasslands are the principal basis of pastoral economies and cultures, affecting their nature as much as the use of the Nile affected Egypt's or as the use of coal and iron affects North American culture. But even modern large-scale mechanized manufacturing may not be wholly or even largely dependent on local deposits of minerals. To take a national example, Japan almost entirely lacks iron ore and petroleum and has only poor and limited domestic supplies of coal. One of Japan's greatest resources is that it is located near the center of the East Asian market and served by cheap water transport. This advantage has helped to make Japanese industrialization possible, while in neighboring China, where coal and iron ore are much more plentiful, industrialization was until recently painfully slow and much smaller in scale. A hundred years ago Japan's location and mineral deposits were the same as now, or better, since much of the mineral reserve has been consumed since then, but there was little or no industrialization, nor was there in mineral-rich North America a hundred and fifty years ago. Human knowledge and technical abilities are clearly as important as location or minerals. Location may not be a usable resource, or may indeed tend to isolate an area and retard economic development, until knowledge, techniques, market, and transport have been developed which make a particular location advantageous.

The biotic resources as a whole are, however, the most basic of all, because human life itself would be impossible without them and because they continue to support us more importantly than anything else which we use. Plants provide directly about 85 per cent of the world's food and indirectly all the rest of it, since all animals and fish depend directly or indirectly on plants. Most fibers are plant material, including many synthetics like rayon, which is made from plant cellulose. Vegetable oils, medicines, dyes, alcohol, building materials, and fuel are only some of the most important of the other plant products which sustain us. Plant and animal products can now more easily be distributed around the earth, and any one area does not itself have to produce them. Because in industrialized economies technological improvements have also made it possible to produce what is needed from plants and animals with only a small part of the total labor force, we may be inadequately aware of our vital dependence on the resources of the biotic world and of the actual bulk of the world economy which their use represents, in terms of area, of numbers of people employed, of goods produced, or of their contribution to our total income and welfare.

The usability of biotic materials depends on the same sorts of factors which govern the use of all resources, however. Knowledge and technical ability are essential, and these differ from place to place and from century to century. Accessibility, distance from market, existing transport means, availability of capital, market preferences and effective demand, and competing sources of equivalent goods also affect the use to which any given biotic or nonbiotic material is put.

Unused "resources"

If a resource is something on which we depend, then materials, advantages, or locations which for any reason are not usable cannot be called resources. Another example may help to emphasize the variety of factors which affect resource usability and the fact that resources are not simply physical phenomena but come into being or become effective only under certain economic, technological, and other conditions.

Iron ore in Brazil

There is a huge quantity of rich iron ore in south-central Brazil, in the province called Minas Gerais (General Mines) just north of Rio de Janeiro (see Map 27 on page 216). It is relatively easily and cheaply mined, and there is a great and increasing demand for iron ore on the world market. But this Brazilian ore was until recently relatively little used, either for export or for Brazilian iron and steel production. How is this apparent puzzle to be explained? Consider the matter of exports first. The market for ore lies almost exclusively in the northern hemisphere. Several other excellent sources of iron ore are nearer than Brazil to the major markets in both physical and economic distance: Venezuela, Cuba, Newfoundland, Labrador, and west Africa, among others, not to mention domestic European or United States ores. In nearly all of these places the ore can be shipped to the market more cheaply. In Brazil the major ore deposit lies about three hundred and fifty miles from the sea in rough highland country. The rich Venezuelan ores are much closer to the sea and can be transported on the navigable Orinoco River. In Newfoundland, Cuba, and west Africa the overland distances from the ore deposits to the sea are also small. These ore sources constitute for the American or European market *intervening opportunity*. Brazil has a cost disadvantage, and does not export significant amounts of ore.

Why was Brazilian ore not used domestically to make iron and steel or to support a general industrial complex? To begin with, Brazil has little coal, and indeed South America as a whole is almost entirely lacking in this vital raw material. The nearest major coal deposit is in the extreme southeastern part of Africa, an impractical distance away. Coal suitable for making fair-quality coke[1] is present in Brazil, but it is not produced in adequate amounts and is also a long distance from the iron ore in the southern tip of this huge country, an expensive haul for such a bulky commodity. Large-scale heavy industrialization in areas without coal is extremely difficult in competition with other areas which have coal or have economic access to it. Coal is important because manufacturing uses such large amounts of it as fuel, as chemical raw material, or as a source of steam power which may also be used to generate electricity. For many of these uses substitutes are available, but they are at present more expensive than coal except in special circumstances. Electricity, for example, can be used to smelt ores and to make particularly high-quality steel, but at a relatively high cost; the same applies to petroleum or to other fuels. Coal is bulky and relatively low in value by weight, so that it is normally not moved very great distances. Coal also loses much of its weight in being reduced to coke, and most of its weight in the manufacturing process as a whole, when it is transformed into a component of iron, steel, or some other product. Raw materials of which this is true (weight-losing raw materials) are usually not transported very far in their raw state; it is obviously more efficient to use them near where they occur. If an area has large supplies of usable coal, it may draw the other necessary raw materials to it for manufacturing in order to minimize total transport or assembly cost. Such an area may therefore support dense populations and large markets.

Brazil's shortage of coal, important as it is, is not the only factor. For other reasons as well, the Brazilian market was until recently limited. The total population was not large in proportion to Brazil's area. The Amazon Basin, which occupies a large part of the country, is covered with dense rain forest and is very sparsely occupied, mainly by subsistence groups. Most of the rest of Brazil is rugged highlands. All of this contributed to the smallness of the market. Capital or surpluses for industrial investment were also scarce; most of the now rapidly growing development of the iron ores for domestic manufacture was originally financed with United States capital. Small markets and limited or high-cost capital generally mean high production costs, because the economies of scale are lacking. Especially in the heavy industries like iron and steel, with their large original investment and fixed costs, the most economic units tend to be large, and, below a certain size, production is too costly to be competitive, either for the local market or for export.

Most of these unfavorable factors in Brazil have now changed. The domestic market is growing in size and in income level, foreign and domestic capital is increasingly available at lower cost, and iron and steel are being made on a significant scale. Other factors may change in future; practicable substitutes for coal may be found, or other technological changes may make it possible for Brazil to support many more people at a higher per capita income level than it now does. The iron ore of Minas Gerais is becoming a resource, however, only as these things happen.

Changing resources and human resources

Since human civilization began, economic, cultural, and technological changes have continually been altering the pattern and extent of resources. Because such change is not uniform over the world, there are pronounced areal differences at any given time in what constitutes a resource. "Resource" is less a physical fact than a cultural, technical, and economic appraisal. This applies not only to minerals but to all of the things on which an economy may depend, including location, climate, soils, biotic resources, and every usable non-human advantage or material. Japan's location or Great Britain's, California's climate or Cuba's, did not become resources until the technical and economic qualities of their respective occupants and the changed spatial relations of these places with other places made their use possible. Once the non-human resources were put to use population increased, created new human resources, and the income level rose proportionately even more. Non-human resources are part of the physical environment. Human labor and skills, varying from place to place and from time to time, may also be considered resources and in many situations may be an area's chief resource. This is obvious enough that it should be unnecessary to stress it, but there is a logical and necessary distinction to be made between people on the one hand and their non-human environment on the other. Attention here can be given to the non-human resources as they are brought into being by varying human cultures, technologies, and economies in distinguishing and employing the usable parts of the environment.

Population densities and resource use

Population densities are understandable only if resources are considered in flexible terms. It has been estimated, for example, that the winter population of interglacial Britain (say in 10,000 BC) was about three hundred total. The same area now supports well over fifty-six million people on an enormously higher level of per capita income or consumption, because more of the potential resources of Britain have been put to work by a technically more advanced as well as a more numerous population. In the course of the industrial revolution Europe's population increased seven times, at the same time increasing its per capita income. Europe's latent resources and physical environment changed very little, but they came to be much more effectively used. The same process took place even more strikingly in the United States, whose population increased fifty-six times between 1790 and 1980[2] as new resources came into being. Obviously, measures of population density or of overpopulation mean nothing unless they can be related to usable resources. Significant measures are not only a matter of the number of people per square mile; the often used term "man-land ratio" may therefore be misleading. The "man-resource ratio" is a clumsier term and may also be misleading unless resources and their use are understood as changeable, but properly understood it is a better measure of the population balance.

China and India, for example, will remain overpopulated, in terms of average per capita income or consumption (the best measure of overpopulation), as long as they make inadequate use of their present and latent resources, which in each country are potentially capable of supporting even more people on a very much higher standard of living than now. Europe, and the rest of the world, were also in a position of relatively low living standards and inadequate use of potential resources two hundred years ago as compared with Europe's present position. The West broke out of this situation primarily by means of an economic and technological revolution, first in trade and agriculture and then in manufacturing, which made it possible to use previously neglected resources and to use others more effectively. The effects and implications of the commercial revolution, or the great increase in exchange, beginning especially with the Western Renaissance, have already been discussed in earlier chapters. The agricultural revolution, beginning in the seventeenth and eighteenth centuries and greatly aided by the development of cheap exchange, was no less important than the subsequent revolution in manufacturing. Improvements in the efficiency and output of agriculture made industrialization possible by freeing labor from the farms, by providing a food surplus to feed cities and industrial workers and to form a source of capital for investment, by raising incomes and widening the effective market, and by stimulating interregional exchange. Improvements included the development or refinement of techniques like crop rotation, the use of clover and other leguminous crops, better plowing and fertilization, improved livestock breeds, improvements in the major crop plants and their wider spread, and new, highly productive crops like potatoes (from America) and turnips, which yielded well even in soils or climates which were poor for grains.

In the manufacturing or power revolution which followed, some new resources were put to work, but, just as important, resources which were already known were used much more effectively or productively. Coal, for example, had been used before in Europe and much earlier in China but not on a large scale as coke to make iron and steel, nor in its raw form to produce electric power, steam power, or synthetics. Steel and machines had been made before, but never so cheaply or in such large quantities. New, more complex, and more productive machines were developed, and new sources of power such as hydroelectricity, steam, and liquid fuels came into use; these were in effect new resources. The commercial, agricultural, and industrial revolutions increased the supply of goods so much that the numbers of people could not only increase tremendously but could also have a much greater per capita share of goods than before.

Population growth

Where industrialization or commercialization has taken place so far, it has been accompanied by a great increase in population (see Table 1 on page 182). But after a time, judging from Western experience, the population tends to stop increasing so fast and may even begin to level off or to decline slightly. The reasons for this are not entirely clear, nor is it certain that population necessarily stops growing in the later stages of industrialization. It was expected to do so in the United States by about 1940, but perhaps as a result of the Second World War population increased rapidly in the United States between 1945 and 1970 and has only recently begun to slow down; no one is quite sure why. Industrialization began earlier in Europe and population growth there had slowed down considerably by 1940, apparently reaching equilibrium in some areas; but after the Second World War, European growth rates rose again. Even during the nine-

Table 1 World population 1650–1980 * (in millions)

	1650	1700	1750	1800	1850	1900	1950	1960	1970	1980
North America	1	1	2	6	26	81	165	199	228	248
Middle America	6	6	5	10	13	25	51	66	86	130
South America	6	6	6	9	20	38	111	140	198	270
Europe (without USSR)	96	106	117	156	214	303	392	427	466	490
Asia	317	362	438	547	793	973	1330	1679	2038	2600
USSR	7	17	27	37	60	126	201	214	244	266
Africa	100	98	95	90	95	120	198	254	352	480
Oceania	2	2	2	2	2	6	13	16	21	28
World total	535	598	698	857	1223	1672	2461	2995	3633	4512

*In most cases, figures for years before 1900 are necessarily rough estimates, as are some of the figures for years after 1900 for areas without adequate census data, such as Africa, Asia, and Latin America.

Sources: United Nations Demographic Yearbook; Encyclopaedia Britannica; W. S. and E. S. Woytinsky, *World Population and Production* (Twentieth Century, New York, 1953); A. M. Carr-Saunders, *World Population* (Oxford University Press, Oxford, 1936); national censuses; author's estimates from other historical materials.

teenth century, the demographic experience in different European countries was apparently contradictory and bore no consistent relation to the timing or scale of industrialization or urbanization. In most of the rest of the world commercialization and industrialization are still relatively recent, and populations are still growing rapidly. Not enough is yet understood about population trends and their causes to predict with any accuracy what may happen to population totals or growth rates in any part of the world. Nevertheless, certain basic facts are reasonably clear.

Death rates are high when food and other goods are scarce and when exchange is limited. High death rates and low life expectancies are characteristic of all economies of scarcity; they also retard economic development, since the economy must support people whose productive life is short or who may die before they reach a productive age. The higher proportion of aged people in most industrial societies presents a relatively smaller but growing economic burden. Increasing the supply of goods and the means of exchange, especially for food and medical services, results in a drop in the death rate. This usually happens relatively quickly, although the death rate may continue to fall more slowly for a long time. But the birth rate continues high, usually for several generations, or it may even rise, as it apparently did in the early stages of industrialization in Europe; this results in a large net increase in the population. Economies of scarcity are also usually characterized by high birth rates which are in approximate balance with the high death rates. It apparently takes time for people to adjust to the changed conditions which follow commercialization and industrialization, to lower death rates, and to the fact that all or most of their children will survive to maturity. It is no longer as necessary to concentrate on simple survival and reproduction as it was under the earlier conditions of scarcity. In western Europe it took about two hundred years from the beginning of the industrial revolution for the birth rate to come down into approximate balance with the low death rate and thus to adjust to conditions of abundance. Most of Africa, Asia, and Latin America are still characterized by relatively high birth and death rates, but death rates have begun to fall rapidly and net population growth rates are hence very high. As commercialization and industrialization progress further, the populations of these areas will probably continue to increase at least as rapidly (see Figure 23). The hope is that the wider and more efficient use of resources which commercialization and industrialization make possible will be able to keep the supply of goods permanently ahead of the number of people, and that eventually the birth rate will fall. But there can be no assurance of this, and there is growing recognition of the dreadful implications of a world population which periodically grows faster than food production. Already the Food and Agriculture Organization of the United Nations estimates that nearly half of the world's present population is seriously undernourished, and that total world food supplies will have to double by AD 2000 merely to maintain the present inadequate nutritional levels. By that time it is estimated that present consumption of energy and of metals will have tripled. For the past

WORLD POPULATION 1968 & 2000
(Area of continents drawn in proportion to population)

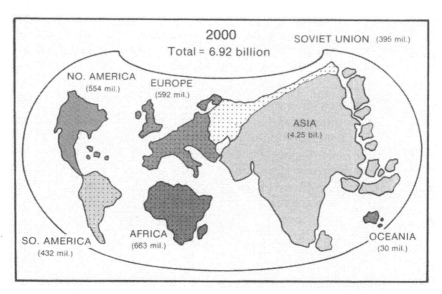

Figure 23 World population, 1968 and 2000

several decades world population has been growing by about 2 per cent per year. Such a rate results in a doubling every thirty-five years.

Other factors in population change

There are other, less well-understood reasons for the growth and decline of populations, apart from the effects of disease and war. Not enough is known about the problem even to say for certain whether industrial-ization is the major factor, although it is clear that populations can increase substantially without industrialization. This happened, for example, in most of Asia as early as the seventeenth and eighteenth cen-turies, before Western-developed techniques were introduced there. The transition from Neolithic to bronze- and iron-age agriculture must have been followed everywhere by large population increases; there is clear evidence of this in China, India, Mesopotamia, Egypt, and Europe. In much of Asia the great population increases of the past hundred or hundred and fifty years, without substantial industrialization, probably resulted at least in part from increased agricultural productivity, commer-cialization and exchange, foreign trade, and the spread of transport facilities and public health services. In Europe the increase in agricultural productivity which preceded the industrial revolution was accompanied by large population increases. All of these pre-industrial or non-industrial increases in population are broadly explainable by increases in the scope and effectiveness of resource use.

Literacy, especially beyond the level of simple ability to read and write, including the interest and ability to read widely and to pursue varied intel-lectual interests, also seems to be related to declining birth rates. Even in some rural or non-industrial areas, relatively high literacy has been accompanied by relatively low birth rates. In more general terms, when people accept the idea that they can to some extent influence their own economic well-being, they are more likely to limit their families in order to provide better for a smaller number of children. Population may also grow less rapidly in the later stages of industrialization because propor-tionately more people live in cities. Urban life may offer fewer incentives for large families than life on the farm, and it has so far generally been true that city birth rates in time become lower than rural birth rates. Some students have suggested that urban life may physiologically affect human fertility. On the other hand, France remains an agricultural country, and most of its people live in towns or on farms rather than in cities; yet France's population leveled off about 1910 and subsequently declined slightly. It is also true that in most American cities birth rates have been higher since 1940 than in most rural areas of the United States, although high urban birth rates may be attributable to large groups of industrial workers whose literacy level is low. In most of the so-called under-developed world as well, urban birth rates are in many cases higher than rural, although it may be reasonable to assume that this is a relatively short-run phenomenon as traditional village attitudes on the part of recent migrants are replaced by urban ones. But in the meantime, the age structure of the urban population favors high fertility, and infant mortality is usually lower than in rural areas.

These complexities are a good illustration of the handicaps of social scientists in attempting final or precise answers or prediction based on cause and effect relationships. They cannot usefully test their theories on mice or guinea pigs, they cannot manipulate people, and they can neither successfully isolate any aspect of human society nor pick out with certainty the particular factors which control the growth of population. They are limited to an analysis of what has actually happened, but for this they often have inadequate information. It is not known, for example, what the exact population of pre-industrial Europe was or its standard of living, let alone the populations of the rest of the world before the present century; even now, reasonably accurate census information does not exist for perhaps half of the world. Recent research does suggest that per capita income or consumption, especially food, in early modern pre-industrial Europe was much higher than had previously been supposed and probably higher than in modern China, India, or most of the so-called underdeveloped world of today; it is certainly unwise to assume that Europe's demographic experience will be reproduced elsewhere. There is also much evidence to suggest that birth control in some form, including delayed marriage, was widely practiced and family sizes were limited in many or even in most areas of the world long before the industrial revolution, in response to changing conditions and varying degrees of economic opportunity for the support of population. It is important to remember that although one development precedes or is associated with another the two are not necessarily causally related. A great many factors affect population growth. Broadly speaking, it is known only that population increases have followed industrialization and have also followed other innovations in resource use, and that in time, usually in association with a high degree of literacy and urbanization, birth rates have tended to decline in industrial societies and populations have tended to level off. Both the increases and the leveling off (if it occurs) probably result at least in part from an increase in the supply and exchange of goods, which is in turn the result of a wider and more efficient use of resources.

Malthus and the "dismal science"

The British economist Thomas Malthus, whose study of population was first published in 1798, argued that it was the natural tendency for population to increase geometrically, but that the production of food and other goods could increase only arithmetically. Sooner or later, therefore, must come a time when the population exceeds the supply of goods and is thus checked from further growth by famine, or by disease and war, which Malthus realized were related in some degree to famine. He urged social and moral restraint as more desirable controls in order to avoid the less pleasant limits which would otherwise automatically operate if the population became too large. This gloomy analysis helped to earn for economics the title of "the dismal science." Malthus was easy to ridicule because in Europe the publication of his book was followed by an enormous increase in exchange and in industrial and agricultural productivity which destroyed the basis of his argument. Not only were more goods produced per capita, but cheap volume transport facilities made possible the distribution of surpluses and thus virtually eliminated the famines which had formerly resulted from regional crop failures. It is also far from sure that Malthus's analysis of pre-industrial Europe was correct, especially in terms of average incomes and of the checks on increases in the population. For much of the rest of the world, however, his analysis may still have some validity, and it can be wholly discarded only when the technically underdeveloped areas undergo the same technological revolution that Europe and North America experienced.

Birth control and living standards

Birth control programs may not always be effective on a large scale in retarding population growth in situations of economic scarcity. Birth rates are high in such situations, in part because survival is emphasized, and they are encouraged to remain high by high death rates. Various methods of family limitation have been known and practiced since at least the time of the ancient Egyptians. But in most non-industrial societies children are especially valued, as additional workers, as security for old age, and as perpetuators of family and individual senses of existence; the childless man or woman is a pitiful figure. Until survival can be assumed, until the economy uses resources effectively enough to guarantee conditions of abundance, until parents can concentrate on providing amply for a few children rather than producing many children in the hope that some will survive, birth control is not likely to be widely practiced even if it is available. This is still to some degree the situation in many so-called underdeveloped economies, although it is gradually changing. In Japan it changed rapidly, after Japan went through a technological and urban-industrial revolution. Economic development is a relative term, and by comparison with the West, Japan is by no means economically or technically retarded. Birth control is widely practiced in Japan; population began to grow less rapidly after about 1955 and may soon level off. The economic margin is wider, and, as in much of the West, children in this urban-industrial-commercial society are more an economic liability than an asset.

Several recently developed techniques of contraception, especially the birth control pill and the intrauterine device (IUD), a simple plastic coil which can be mass produced very cheaply and is reasonably reliable, may represent tremendous positive contributions to human welfare. In much of the world, the impoverishing pressures of population are clear enough to many individuals and to planners that a cheap and simple means of controlling conception might be widely adopted. Any projection of current population growth rates produces an alarming result even within the next couple of decades (see Figure 24). Malthus may yet be proved right. If the present perilously low living standards of most of the world are to be preserved, let alone increased, means must be found to control the number of people as well as to increase the supply of goods before mankind breeds itself into disaster.

Per capita income levels are seldom strictly comparable between different cultural and economic areas, however. They do provide the best and

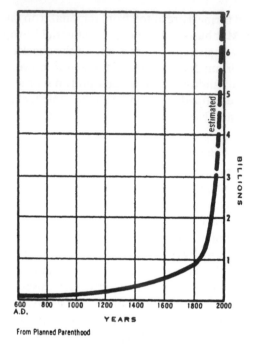

From Planned Parenthood

Figure 24 World population growth, AD 600–2000

most convenient single measure of economic development, resource use, and population balance, assuming the necessary data are available with sufficient accuracy, which is not always the case.[2] But per capita incomes may conceal major differences in other respects. Different cultures and economies attach differing values to economic goods and to other matters. Subsistence economies may have an acceptable standard of living in their own terms which would seem completely submarginal to individuals in a commercial economy. Such differences may also exist between different subsistence or different commercial economies. Some goods, services, or wants, especially in subsistence economies, are difficult to evaluate in money terms or to convert into per capita incomes. Most Western societies, particularly the United States, have until very recently believed firmly in the concept of progress and measure it largely in economic terms, or in terms of the number and variety of economic wants which can be developed and can be satisfied: the more realizable wants for goods and services, and the more goods and services consumed, the better. In many or most other societies such attitudes are far less prominent, are wholly absent or are even contradicted by other and non-economic values, although societies may acquire some of these essentially Western and often destructive values and standards as they industrialize. Affluence beyond reasonable needs may create more problems than benefits; emphasis on *individual* affluence certainly hampers the development of group or public welfare. The quality of life cannot be measured by the quantity of consumption. In any case, abundance and scarcity are clearly relative terms, as is economic development, and they will continue to be so even if and when industrialization and commercialization are spread more widely over the world. The process of what is loosely called economic development involves certain common problems, however, and it is worth examining in those terms. One of the most basic of these common problems is the development of resource use, including the harnessing of energy. The following chapter surveys economic development on a global basis and considers the special place of energy in the pattern of resource use.

NOTES

1. Coal is reduced to coke for making iron and steel. By no means is all coal suitable for making metallurgical coke. The coal must have the right chemical and structural properties, ash content, heating value, volatility, and other characteristics.
2. The United States population totalled 3.9 million in 1790, 131 million in 1940, 150 in 1950, 179 in 1960, 212 in 1970, and 221 in 1980.
3. Completely accurate figures for total income or production are almost never obtainable for any area. Equally important is the fact that income figures may not be comparable. It is difficult, for example, to compare a pound of bananas as income in Costa Rica with a pound of furs as income in Finland. To convert each item into a monetary equivalent conceals real differences in value or desirability in each area. Wants and consumption habits vary widely, for many different reasons. Computation of total or per capita income is usually easiest in commercial economies, where most things are readily measured in money terms, and where relatively more accurate data are available. Such figures are seldom comparable with figures obtained from subsistence economies or from situations where different values predominate.

14 ENERGY AND DEVELOPMENT

Have we been successful in our use of the earth, and in the human systems which we have constructed? Until recently, most people would have had little difficulty in answering yes to both questions, even those in the poor ("developing") parts of the world, because even there the material conditions of life for most people, as for most in the richer countries, have improved, especially during more recent years. In such terms, we have made progress since the emergence of *Homo sapiens*, since Neolithic times, classical times, and medieval times. Life expectancy, perhaps the best single measure of material welfare, has risen nearly everywhere; famine, malnutrition, and disease are overall less severe constraints on human life than they used to be. We have largely got rid of slavery. The industrial revolution has made the means of relative abundance potentially available to all. But increasingly since about 1950, there has been a well-based and growing concern about two disturbing problems (leaving aside the threat of nuclear war) which must fundamentally qualify any declaration of human "success." The almost religious belief in continuous "progress" which was characteristic especially of Western industrializing societies in the nineteenth and early twentieth centuries has yielded to a more sober awareness that our record is flawed and our future cloudy.

First, we have been dangerously profligate in our use of finite energy sources, and in that process have done terrible damage, through pollution, to the environment which must sustain us. *Second*, we have notably failed to distribute the abundance which we have the means to produce. The poorer countries – three-quarters of the world's people – are being left farther and farther behind as the rich countries grow ever richer, while growth in the rest of the world is much slower or even static in per capita terms, since population there continues to increase rapidly. The problems are interrelated, since it is the mushrooming use of energy which has made industrialization and wealth possible for the rich, in many ways at the expense of the poor in their role as suppliers of raw materials for other peoples' development.

Energy use

Energy has become a buzz word, but it is indeed the basis of all life and always has been, as well as being the key to economic development. We have become newly sensitive to it, especially in North America, because we are so heavily dependent on its over-use, at a time when most present world fuel sources have become scarce and expensive. They have become so, of course, because we have used them so prodigally. Americans especially are deeply addicted, as if to a drug, to energy use on a gargantuan scale and in an endless variety of ways. While it is both easy and appropriate to criticize this, and to try at least to reduce the addiction, it is an understandable outgrowth of the American situation. We did not start out that way; what gave the thirteen colonies and the early United States their distinctiveness and helped to spark their struggle for independence was the contrast they saw between themselves as poor, simple, and frugal, and Britain and Europe as rich, extravagant, sophisticated, and corrupt. The industrialization of the United States, based on superabundant domestic resources including energy sources, changed all that. *We* became rich and powerful, at the head of the development process. It seems regrettably to be part of human nature that wealth and power corrupt, but this is something which has applied to everyone, not just to North Americans.

What has made North American use of energy and other resources such an extreme case is the unrivalled wealth of our resource base, discovered and put to use only since about 1850 (with techniques mainly imported from Britain, where industrialization was born). Equally significant, these riches were shared by a relatively small population, even now only about 5 per cent of the world total. To support its industrialization drive, North America had about a third of the world's reserves of coal, about a fifth of its iron ore, a fifth of its oil, and superabundant reserves of most other minerals, plus the world's largest area of high-productivity farm and forest land and fish resources. Prodigal use of this wealth developed naturally, however shortsightedly. We have been used to abundance and have long forgotten our earlier emphasis on frugal management, a skill which it now seems clear we must re-learn if we are to survive. The party lasted about a century – from 1850 to 1950 – but it is over now. Energy and most other industrial resources are dwindling, even in North America, and we are only beginning to realize the harm, perhaps irreparable, we have done to the environment in the course of using them.

The United States alone, with less than 5 per cent of world population, accounts for over 30 per cent of world energy consumption, including half of all natural gas use, 30 per cent of all oil, and 20 per cent of all coal.

Americans use more energy for air conditioning alone than is used by the one billion people of China for all purposes. Per capita energy use in the United States is about 250,000 kilo-calories per day; the average for the other industrialized nations is about 125,000, and for the world as a whole about 30,000. Despite our rich domestic resources of oil, we now import about half of what we use, mainly from the developing (poor) world. Much of United States energy use is wasted: over-heating in winter, over-cooling in summer, heavy, bulky and inefficient cars and profligate use of them, throw-away containers and packaging (much of it, including plastics, made from oil and/or using much energy in its manufacture), powerized gadgets, dishwashers, driers, hair driers, electric carving knives, and endless other symptoms of a wasteful society which has forgotten how to conserve or to be efficient in the use of anything scarce, an experience lost in the party days from 1850 to 1950.

Our throw-away culture is not only wasteful but dangerously polluting. Nothing can really be thrown away, merely dumped somewhere, or converted (e.g., by burning) into something else; both produce problems whose results, as in the case of toxic chemical wastes, are harmful or even fatal to life. The private car, probably a wrong turn we took two or three generations ago but which is hard to reverse now, may be the most wasteful and destructive aspect of American energy use, especially since we have, like so many other things, carried it to an extreme. There is nearly one car for every two Americans, far more than anywhere else in the world; the car has become almost like an extension of self. To make the problem more difficult to solve, a very large proportion of the United States labor force is employed in automobile-related industries. It is probably not reasonable to try to cure the American system completely of its addiction to the private car, but we need to control and reduce the worst of the problems it causes: energy waste and pollution. Europeans find most of our cars laughable – huge, heavy, ungainly things loaded with powered "extras" (including air conditioners) – and our recklessly wasteful use of them equally so. Present car engines are notoriously inefficient, the bigger the more so, wasting some 80 per cent of the potential energy in gasoline. Something can be done to improve this, and a few such changes are already beginning, but in the meantime automobiles account for over half of United States oil consumption, 70 per cent of its rubber consumption, and produce about 80 per cent of the air pollution as well as the urban noise in most United States cities. Is this really "progress"? What would Thomas Jefferson think of it?

The energy budget and the energy cycle

Let us survey briefly the global energy balance and energy cycles, and then consider aspects of the current "energy crisis," leading into a discussion of the closely related problem of economic development. All energy on the earth comes from the sun; without sunlight, all terrestrial life would end. Three days' worth of direct sunlight on the earth is equivalent in energy to all known reserves of coal, oil, and natural gas. These *fossil fuels*, as they are called, are in effect captured sunshine in the first place, the product of vegetation which eons ago stored sunlight and then was highly concentrated and subjected to subterranean heat and pressure to form hydrocarbons. In a more current sense, sunlight provides the energy for all plant growth, which in turn is essential to the maintenance of all animal and human life. The sun is also the engine, through its heat, of atmospheric and oceanic circulation patterns which provide water, through precipitation and run-off (surface streams), the other essential element of all life.

Solar heating of the atmosphere and of the surface of land and sea evaporates water, raises it into the atmosphere, and produces pressure, wind, and current systems which circulate heat and water. As a result of temperature and pressure differences and convergences, fuelled by solar energy, evaporated water is precipitated out of the atmosphere as rain, snow, sleet, hail, or dew (see Chapter 8). Some of this moisture is absorbed and used by plants to sustain their growth, some is transpired (breathed) back into the atmosphere by plants (especially by trees), some soaks into the ground and sustains ground-water reserves in the water table, some is evaporated directly from the surface, and some runs off into lakes or streams which eventually feed into the oceans as the major collecting basins. Streams carry salts and other minerals in solution, picked up from the rocks and soils through which they flow, and deposit these in the oceans, where they have built up into salinity because of solar evaporation, which leaves the salts and minerals behind. (Streams and most lakes are too small to be affected in this way, and are continually recharged with fresh water from precipitation and run-off.)

About 30 per cent of incoming solar energy is reflected back into space by clouds, dust, and the earth's surface, and about 70 per cent is absorbed, of which about 25 per cent goes to the evaporation of water and about 45 per cent to heating the atmosphere and the surface of the earth. Most of this heat is then dissipated outward by radiation. Only a very small part – less than 0.003 per cent – is captured by plants and hence enters the food chain. Even so, the total solar energy captured and converted to chemical energy by plants alone is *annually* over *four hundred times* the total fossil fuel (including nuclear) energy used by people each year. A small part of the solar energy stored in plants is then transferred along the food chain to animal and human life. Each step in the food chain, however, involves the loss of most of that energy as it moves from the sun to plants to animals to humans. A thousand calories of solar energy may produce, for example, 100 calories of grain, which in turn may yield 10 calories of energy to an animal or human consumer. If however the grain is used to feed animals for human consumption, involving another step in the food chain, the 100 calories of grain may produce 10 calories of meat, which in turn yields only 1 calorie of energy to a human consumer.

As incomes have risen following industrialization, people have tended to substitute more and more meat for grain in their diets. This is clearly seen in the contrasts now between the poorer and richer countries. According to 1975 United Nations figures, per capita meat consumption in the "developing" countries averaged 10–20 pounds (5–10 kilos) per year, but in the "developed" countries the average was 200–250 pounds (100–125 kilos). The United States is again an extreme example, where nearly 90 per cent of the grain consumed is fed to animals to provide meat for the table. In addition, there is massive use of fossil fuel energy (machines and fertilizers) to grow the grain and to manage, transport, and store or freeze animals and meat. Range-fed animals, which graze on natural vegetation (mainly grasses) are of course more efficient sources of meat but their meat is leaner and tougher.

Energy and agriculture

Agriculture as a whole is a big user of energy, especially in the industrialized countries where machines, fossil fuels, chemical fertilizers, and pesticides are available and affordable. Human and animal energy are dominant in the agricultural systems of the poorer countries, and in this sense they are far more energy-efficient. But there is a genuine dilemma here. The poor countries face a continual threat of food shortages as their populations continue to grow relatively rapidly, for the reasons examined in the preceding chapter, and while they remain in the early stages of adequate industrialization. There is a pressing need to increase food output as rapidly as possible. There are basically three ways to do this: increased irrigation (which requires steel, cement, machines, and fuel for dams, canals, wells, pipes, and pumps); improved, higher-yielding varieties of crop plants (which requires technologically-backed research and development); and increased use of fertilizers and pesticides (which requires a large and sophisticated industrial base). All of these things represent greatly increased energy use, especially the production of fertilizers and pesticides, many of which are also made from oil or coal.

What has been called the Green Revolution in much of the poor world has rested on a combination of these three factors. Higher-yielding strains of wheat and rice were developed through research, and beginning in the early 1960s were widely disseminated. Their yields cannot however be kept high without ample supplies of both water and fertilizer, far more of both than most traditional (pre-industrial) agricultural systems could provide except in small areas. There thus remains great pressure to increase irrigation and fertilizer output, which means more and more energy use. Pesticides are also badly needed. Pests, plant diseases, and spoilage are estimated to destroy *over a third* of world food each year, most of it in the poorer countries where controls and storage facilities are poor or limited. Much of this loss is preventable, but at a cost in both the energy to produce pesticides and better storage, and in the environmental damage which most chemical pesticides cause. This is a worrying matter, since like almost all environmental damage, especially from chemicals, we do not yet know how serious, long-lasting, or even permanent this damage may be. But for the poorer countries, the food problem is immediate and urgent; they cannot afford to waste time or to temporize. We can only hope that safer or less destructive chemicals, or natural, biological controls, can be developed and used which can do most of the job without the heavy risks.

So far, the Green Revolution has been a success in that yields and total output, especially from the major grains, have risen rapidly in most of the poorer countries (as they did earlier and continue to do in the richer countries), wherever adequate water and fertilizer have been available. But it is clearly a conditional growth, wholly dependent on the maintenance and increase of those two factors. As energy costs continue to rise, agriculture everywhere, and the Green Revolution itself, will encounter increasing problems, worst in the poorest countries where the burden of higher energy costs is of course greater. India, for example, must pay the same price for the oil it must import as the United States pays, even though India uses that oil far more efficiently and for much more essential purposes. The socio-economic aspects of the Green Revolution have also often been unfortunate, since in most poorer countries only the relatively few better-off farmers can afford the additional costs of new irrigation and fertilizer inputs. In many countries this has resulted in a small new class of rich farmers (since the increased output is very profitable) and a relative impoverishment of most of the others, many of whom have been unable to compete, have lost their land, and must work as hired laborers, or migrate to the already overcrowded cities.

The oceans: resource or sewer?

The oceans have traditionally been seen as a boundless source of food in the form of fish, even though only a very small part of the human diet – perhaps 5 per cent – over the world as a whole comes from the sea. Traditional (pre-industrial) fishing was able to tap only a tiny fraction of the fish resources, and the lack of long-distance, low-cost transport and of preserving facilities greatly limited the use of fish as food. In the present century, fishing has become highly mechanized over much of the world, with large powered vessels, huge nets hauled in by power winches, icing, freezing, or canning on board in *factory ships* which range over all the oceans, and fast distribution on land. Fish schools are located by sonar and helicopter. All of this uses enormous amounts of energy, and the price of fish has risen accordingly. This has also produced what is now clearly revealed as *over-fishing*. Newly effective techniques of finding and catching fish have greatly decreased the total stock and may have pressed many of the most popular species close to extinction, or below the point where they can recover their former numbers. Every year more effort and more energy use is invested in increasingly mechanized fishing. The total catch rose from 1950 to 1976, but since then has sharply levelled off, a clear warning that we have damaged, perhaps irreparably, the regenerative base of this important resource. Part of the reason is also that we have so severely polluted the major breeding grounds of many fish, and especially of shellfish, in the estuaries and coastal zones with urban and industrial wastes that many of them have become dead areas. Man-made pollution has also affected the oceans as a whole, but again we do not know how severely or with what long-term results. Meanwhile, mechanized fleets threaten to extinguish whales (despite efforts at international control), perhaps the most impressive form of life on the earth, and have squeezed out or greatly reduced the basis for the remaining traditional-style fishermen in the poorer countries.

Our treatment of the oceans does not make a pretty story. We have used them as a sewer, killed off much of their life for our own profit, and expended huge amounts of scarce energy in the course of that destructive hunt. Efforts to control over-fishing and pollution through an international law of the sea have so far achieved only small results; individual countries are unwilling to compromise their own sovereignty even in an effort to promote the common good. We still know disturbingly little about the oceans and the life they contain; it makes sense to be careful how we treat them, since we can at least be certain not only that they constitute an important resource in themselves but that they also play a critical role in the global balance of energy and life. *Plankton*, minute animal and plant life in scattered patches on and near the surface of the sea, a major source of food for fish and whales, is a possible source of food for humans too. But it is widely diffused and at present it would cost too much in scarce energy, time, and labor to harvest it to make it competitive with other food sources. *Aquaculture*, the raising of fish, shellfish, and edible aquatic plants in enclosed shallow coastal waters and in inland freshwater ponds, is far more promising. This is an extremely efficient means of food production, long practiced especially in Asia. It should be greatly expanded, but is jeopardized in many areas by uncontrolled pollution especially of tidal lands, bays, and estuaries. The oceans may also become more important as sources of minerals or fuels, especially as these become scarcer from terrestrial deposits and costs accordingly rise. Sixty-four of the 92 chemical elements are present in sea water, most of them in very low concentrations except for salt and magnesium. This means that too much energy is required to extract them to make it pay, at least for the present. Mining the sea bed for minerals is also very expensive, probably prohibitively so (barring new technology) except for oil in shallow coastal areas and possibly the extraction of some nodules of iron, magnesium, and copper which occur in certain places on the deep sea bed. The chief task now is to prevent the sickness or death of the oceans through pollution and over-fishing, and to preserve them as vital living ecosystems.

Fuels and power

While food and the interlocking systems of living organisms are the most basic aspects of the energy cycle, industrialization has concentrated attention on fuels and sources of power, the essence of industrialization. Oil has now overtaken coal, providing about 37 per cent of world fuel energy, with coal accounting for about 26 per cent, wood, dung, and crop residues for about 20, natural gas for about 16, and hydroelectricity for about 1 per cent. (For the United States alone the figures are oil 47 per cent, natural gas 26, coal 20, hydro 3, and nuclear power 3.) Of these, only the organic fuels (wood, dung, crop residues) and hydro power are *renewable* or *flow resources*. This does not mean either that they are free or that we can safely use them as much as we like, since renewal or regeneration takes time. But careful use can be extended more or less indefinitely, unlike the fossil fuels, all of which will sooner or later be exhausted.

Unfortunately human pressures on the renewable resources have for some time now been so heavy that adequate time has not been allowed for regeneration; wood especially is becoming critically scarce almost everywhere in the world, as forests and even scrub areas and twigs are cut much faster than growth replaces them. Wood is of course used for many purposes besides fuel, which helps to produce over-cutting, scarcity, and continually rising prices. The boom in wood-burning stoves for domestic heating in the United States as one aspect of the "fuel crisis" from about 1977 has already more than doubled the price of firewood, increased the *cut-to-growth ratio*, and threatens to damage or destroy woodlands within a day's drive of most United States cities. In the dry world of north Africa and the Middle East and in dry South Asia animal dung has long been the dominant domestic fuel, in the absence of forests, but rising population pressures have also made it scarce and expensive. Burning wood, dung, or any other organic material, none of which are very efficient as fuels in relation to their weight or bulk, also produces atmospheric pollution. This has already added to existing air problems in many American cities.

Hydroelectric power is clean and non-polluting, at least in its use, and represents perhaps the ideal *flow resource*. Unfortunately, most hydro plants must depend on a reservoir behind a dam where water is stored so that it can flow continuously in a relatively constant volume to turn the turbines. The reservoir is progressively filled with silt deposited by the river. All rivers carry silt in suspension, but when their velocity slows, as when they empty into a lake, the sea, or a reservoir, their capacity to hold silt in suspension falls off sharply and the silt is dropped. This is readily observable in the continuous formation and extension of deltas at river mouths or the build-up of silt in estuaries. The *average* life of a power dam reservoir is estimated at between thirty and fifty years, varying of course according to the silt content of the river which has been dammed. A silted reservoir can be dredged and re-used; settlement basins to extract silt upstream from the main reservoir can be built. But all this imposes heavy extra costs, and it is often cheaper to abandon the site, despite the heavy original cost of the dam. As pointed out in Chapter 11, hydro power is far from free, and far from problem-free even once it is built. For most areas of the world, it remains more expensive than power from fossil fuels – which is why it is not used more than it is, and why we cannot depend on it to solve the "energy crisis." Unfortunately, cost problems continue largely to rule out what once seemed a very promising source of water power, the tides. Here there should be no troubles with silting, no need for a reservoir, and no problems with seasonal fluctuations in flow. The energy locked in the tides is truly immense, but in part for that reason the costs of constructing a dam or turbines to trap that energy have so far proved too high to make it pay in competition with other sources of energy at present prices. Regrettably, the same limitations appear to apply to projects aiming to capture some of the enormous kinetic (moving) energy involved in simple wave action in the sea. For the present at least, the sea is just too powerful for human efforts to manipulate it at bearable cost.

There are newer projects designed to make use of ocean temperature differences *(thermal gradients)*, where the range from the surface to a depth of a thousand feet is as much as 30° to 40° Fahrenheit (17–22° Centigrade). The heat from warmer surface water would be used to vaporize liquid ammonia under pressure, which would power a turbine to generate electricity; part of the power produced would be used to draw colder water from below to cool condensers for converting the vapor back to liquid. Such a project is fully practicable, but unfortunately there are only a few suitable sites with adequate thermal gradients, and the cost problems of transmitting the power to where it would be used remain. So far, this is not a practical or competitive source of energy. Energy from naturally occurring hot springs is already being used in a number of small scattered areas (New Zealand, Japan, Italy, Iceland) where there is sub-terranean or near-surface volcanic activity. This can be fully cost competitive, and in some areas deeper shafts can be sunk to tap heat at greater depths, usually to make steam for electric turbines. But total practicable sites are few, and *geothermal* power is not likely to become a major source of energy for most of the world.

Wind power has of course long been used on a small scale, to drive ships and, most importantly, windmills. Recent experiments with larger scale windmills have so far been mixed; many of those built have not been fully efficient or reliable, and in net terms expensive as alternative sources of power. There is much competitive scope for smaller windmills, on a household or community basis, and perhaps for a revival of sail-powered ships using machinery only to set and alter sails and to load and unload. But wind power is unlikely to have a large share in total energy use, or for industrial needs. Methane (biogas) has attracted attention in recent years as an alternative fuel, especially suited to poor agricultural countries since it can be made easily and naturally from trapping the gases released in the decomposition of any organic matter, including crop residues, plants, grass, and human and animal wastes. China and India in particular have had good success with small biogas plants, providing fuel for domestic heating and cooking and for some small-scale industry and transport; biogas can also be used to power internal combustion engines. However, it seems practical or competitive on only a small scale, since so much energy – mostly human labor – must be invested in collecting and composting the materials for such a relatively small output of fuel.

There are far more promising possibilities in deriving alcohol, an excellent fuel, from plants, to be used straight or mixed with gasoline. Alcohol can be made from most plants, including grains and potatoes, and from plant residues. Already in the United States many stations sell "Gasohol," gasoline with a 10–15 per cent alcohol content, at a price comparable with straight gasoline. Alcohol is now made primarily from grains and other food crops, but it could be made from other plants grown especially for it, in some cases on land less well suited to food crops, and at a cost increasingly competitive with oil-based fuels as oil prices rise. Some plants, notably the *Euphorbia* group, produce organic hydrocarbons which are structurally the same as those concentrated and fossilized in oil or coal. These could also be grown on a large scale as a source of high-grade fuel which will soon be cheaper than oil, although the amounts necessary to replace oil or coal completely would be huge and would probably be beyond the capacity of even an expanded agricultural system to produce. Another promising development is the use of urban wastes as energy, especially in throw-away cultures like the United States where we are literally drowning in our own garbage. Practically all these wastes can and should be used; non-organics (glass and metal) can be sorted out and re-used; organic materials can be converted into high-grade fertilizer (as in China) or burned in furnaces designed to minimize atmospheric pollution and used to produce power through steam turbines. In Japan, where industrial and household wastes are also a major problem in that very crowded country, the city of Tokyo, among other Japanese cities, has developed a system of disposal which does all of these things at high efficiency, and at a cost which is overall less than continued random dumping, including the revenues from power production. As with other Japanese successes in controlling pollution, this example needs to be followed by the rest of the industrialized world.

Solutions

As the cost of fossil fuels inevitably continues to rise with the double pinch of scarcity and increasing demand, and especially as industrialization progresses in the "developing" world, most of the alternative and renewable sources of energy will become more and more economically practicable. They have been starved for research and development funding and have thus not been given a full and fair trial; that is now beginning belatedly to change, and we may expect somewhat more rapid progress. In the United

States the great bulk of research and development money continues however to go into the fossil fuels, including nuclear power and conversion projects to substitute coal and coal derivatives for oil. This seems extremely short-sighted, but is again a reflection of America's relative wealth in coal and in high technology, plus its insistent appetite for more energy NOW, without adequate thought to longer-run consequences. Of all the alternative sources of energy, by far the most promising is the most basic, sunlight, but discouragingly little research and development have so far been invested in devising improved technicians to harness it. Nevertheless, it seems reasonably clear that only solar energy can save us from the disaster involved in continued commitment to fossil fuels: mounting shortages, galloping costs, and fearful, perhaps lethal, damage to the environment. Current pressures to substitute coal for oil are especially worrying. Coal is the greatest single polluter, not only of the atmosphere through dust (particulates) and noxious gases but of water too, directly and in the form of the many chemicals (and their wastes) made from coal. The sulphur released into the air from burning coal is poisonous, and is also brought down in solution to the ground by rainfall, producing an *acid rain* which erodes stone and metal, kills plants, and poisons animals and humans.

The development of solar energy is still in its infancy, but that can also mean that its great growth lies ahead as a promise. Present techniques for capturing energy from sunlight are simple and primitive, most of them able to extract a miniscule part of the heat and to trap it behind glass, as in any greenhouse. This heat can be stored only in small amounts and for short periods, usually in pipes, cans, or barrels of water, which holds heat reasonably well. With an added expenditure of energy, pumps can circulate the heat stored in water, or fans distribute warmed air. Parabolic mirrors and lenses, especially where rotated to follow the sun, can help to concentrate sunlight for more heat extraction, but at further expense. These techniques have been known and used for many centuries on a small scale. In the last few years their use has greatly increased because of rising energy costs, and some small improvements have been made, especially in water storage of solar heat and in insulation to retain it at night or during cloudy periods. Captured sunlight can provide all or almost all of domestic heating needs in most parts of the world, and at a cost already fully competitive with heat from fossil fuels or hydro power. However, domestic heating is only a small part of total energy consumption. New technology is needed to convert solar heat to usable energy more directly, more efficiently, and on a much larger scale. We may now be on the threshold of such a break-through, perhaps by means of *silicon cells (photovoltaic cells)* which generate electricity when exposed to sunlight. Silicon is widely and cheaply available – it is for example the principal constituent of sand; the problem has been to develop techniques for processing it in a form which can then be sliced into thin wafers to make silicon cells. This may soon be possible cheaply enough to make solar-generated electric power as cheap or cheaper than power from coal, oil, or nuclear plants. If so, it will be an enormous step forward, not only because it can ease or solve the energy crisis we have brought on ourselves, but because it can also reduce or eliminate the worst forms of pollution, which come from the use of fossil fuels.

There are some other signs of hope. Americans, the largest and most wasteful consumers of energy, have begun to reduce somewhat their total consumption of oil (although much of this has unfortunately resulted from conversion to even worse polluting coal), and at least their per capita consumption of energy and of most metals, which require huge amounts of energy to produce. These are only the smallest beginnings, and we still have a long way to go. Total energy consumption must somehow be reduced and kept within reasonable bounds, starting with the United States, the most extreme and wasteful consumer. Meanwhile there are increasing efforts to produce more oil, by getting more of it out of wells through pumping water or steam into them, by more intensive search for new oil fields, including those on the continental shelves under water, and by better techniques to process *oil shales* or *tar sands*, in addition to producing oil from coal. Oil shales, a rock formation containing hydrocarbons, are very widespread in many parts of the world, including the United States; oil produced from them is still considerably more expensive than oil from wells, and mining the oil shales also scars the landscape and yields huge amounts of waste material for disposal. Tar sands, which also contain oil, are very extensive too, but so far even more expensive as a source of petroleum, although more cost-efficient ways of producing oil from both sources may be developed and could help to postpone the "oil crisis" we have made for ourselves.

Nuclear energy

Nuclear power has been left for discussion last, since it remains probably the most problem-ridden of all sources of energy. At one time it seemed that *nuclear fission*, the same process involved in the atom and hydrogen bombs, might provide reasonably safe, cost-efficient, and boundless energy. Accumulated experience since the first nuclear explosion in 1945 has increasingly shown that nuclear power carries the risk of catastrophic damage to all life, and that so far it remains significantly more expensive than power from other fossil fuels, probably also from many of the renewable sources discussed above. Uranium, thorium, and plutonium, the principal raw materials used in nuclear fission, are non-renewable and could easily be exhausted. But the major worry is the lethal effects of excessive radiation, devastatingly demonstrated at Hiroshima and Nagasaki where the United States dropped the first atom bombs in the closing days of World War II but where the longer-term damage to plant, human, and animal life is still unfolding in the form of genetic defects plus increased incidence of cancer.

The earth receives continuous low-level radiation from cosmic rays through the atmosphere; this has now been augmented by nuclear fall-out everywhere and by X-ray radiation received by medical and dental patients. The evidence so far is disturbingly unclear on the long-term dangers from this cumulative increase in low-level radiation, but it now seems likely that the standards set for "safe" levels are too high. There are two major worries in this connection about nuclear power: the risk of accidental release of radiation, including the possibilities of an uncontrolled nuclear reaction which would literally melt the entire plant; and the still unsolved problem of the disposal of the radioactive wastes which are produced even without any accident. There is also the risk of leakage in shipment of raw materials, components, and wastes. As already pointed out, nothing can really be thrown away, and radioactive wastes may remain lethal for hundreds or even thousands of years. Underground storage has already proved far from safe; burial in the sea bed raises similar problems, especially given the corrosive action of sea water; the same applies to projection into space, with the additional drawback of high cost. Basically, there is no fully safe method of disposal for nuclear wastes. Most accidents can be prevented by careful management, monitoring, back-up systems, and fail-safe devices. But "most" is not "all," and even one would be devastating. A partial or total *melt-down* occurred in the 1970s at a nuclear plant in Siberia, leaving a very large area around it lifeless and uninhabitable. The near melt-down of a nuclear plant at Three Mile Island in Pennsylvania in 1979 threatened to produce similar results over an area inhabited by ten to thirty million people. More stringent safety regulations have followed these experiences, but they serve to underline ever-present and frightening risks.

New nuclear plants will probably continue to be built, but more slowly and carefully (hence more expensively) and on a smaller total scale than once was projected. Nuclear fission does not seem to offer an adequate solution to energy shortages, and will probably never account for more than a small part of total energy production. Cost figures, already high, do not include the heavy costs of waste disposal, an unsolved problem which must somehow be met. *Nuclear fusion*, a different process using radioactive elements, can potentially produce much more energy from the same amount of mass, but has not yet proved possible on a sustained and practical basis. It is somewhat less risky than nuclear fission, but far from risk-free, and the problem of waste disposal is essentially the same. In addition, nuclear fusion uses enormous amounts of energy to fuel the process, so that the net yield is reduced – and the costs increased. For the present we are stuck with oil, coal, and natural gas as the major sources of energy. Natural gas is at least relatively clean-burning, but supplies are finite and are rapidly being used up. What is needed is a combination of reduced energy use on a more rational basis and a major technological drive to develop renewable and non-polluting sources of energy. The record of technological change during the past century, and especially in recent decades, suggests that this is not only possible but likely.

Development and the poor world

If and as new energy sources become cheaply available, they could make an important contribution to the economic development of the poor world, and perhaps thus to some evening out of the present disturbing and unhealthy gap between rich and poor areas. The terms "developing" and "Third World" are often used as euphemisms for the poorer countries, "developed" and "First and Second Worlds" for the richer ones, non-Communist and Communist respectively. The rich countries thus set the standard of relative affluence, based on industrialization, and they are still increasing their wealth and power more rapidly than the poorer countries so that the gap between them continues to widen. Rich and poor are of course relative terms, and they are always changing. China and India, by far the largest of the currently poor countries, were far richer and technologically more advanced than Europe until perhaps as late as 1700 – in China perhaps 1800. They have fallen behind *relatively* as indus-trialization and technological change took place far more rapidly in the West after about 1850. As scarce resources, including energy, grow scarcer, the rich countries use their power and wealth to control and use a larger share, in rivalry with each other. Even apart from matters of social justice, this is a worrying trend which clearly threatens the peace of the world.

The gap between rich and poor countries is graphically shown in Maps 23, 24, and 25. Per capita income is notoriously hard to measure accur-ately or meaningfully, but is widely used as a key to level of development. Life expectancy is of course the bottom line, with a further and revealing check through the rate of infant mortality, since malnutrition and disease bear hardest on the very young. (Note that the United States does not compare well in either respect with many of the other rich countries, a reflection of our disgraceful inability to provide adequately for our own poor – a quarter of our population – and perhaps a reflection also of improper diet, *over*-eating, and stressful but physically soft life styles with inadequate exercise – the private car and other gadgets again! – on the part of the rest of us.)

The picture of the world which these maps offer is a grim one, not because there has been no progress toward improved welfare in the poor countries but because material welfare has grown and continues to grow faster in the rich ones. Part of this is due to higher rates of population increase in the poorer countries so that per capita shares grow only slowly. But the world is one; all parts of it and all of its people are inex-tricably linked. We share a common environment, a common set of resources, a common stake in world peace, and a common humanity. We are intimately involved with one another through trade and exchange, of commodities, ideas, institutions, culture, and technology, more so all the time – and also through tensions, rivalries, and conflicts. Pollution in one area rapidly spreads to others, and in the atmosphere or the oceans is circulated to all parts of the globe. The originally Western model of con-

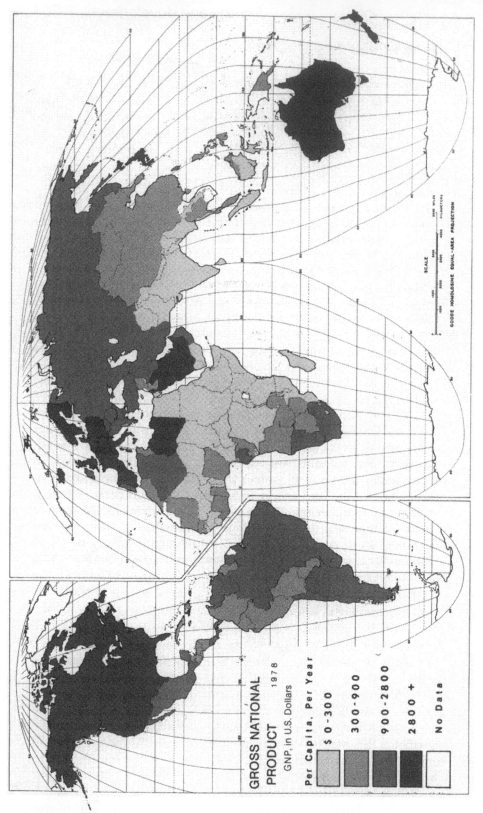

Map 23 World per capita income. *Data are for the late 1970s, slightly different years for different countries.*

Map 24 World life expectancy at birth. *Data are for the late 1970s, slightly different years for different countries.*

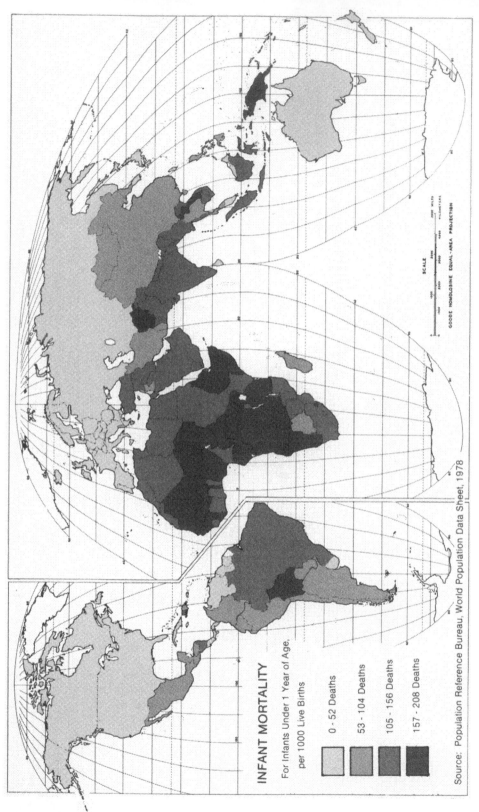

INFANT MORTALITY

For Infants Under 1 Year of Age,
per 1000 Live Births

☐ 0 - 52 Deaths

☐ 53 - 104 Deaths

☐ 105 - 156 Deaths

☐ 157 - 208 Deaths

Source: Population Reference Bureau, World Population Data Sheet, 1978

SCALE

GOODE HOMOLOSINE EQUAL-AREA PROJECTION

Map 25 World infant mortality. *Data are for the late 1970s, slightly different years for different countries.*

centrated industrialization dependent on fossil fuels is badly flawed, but it has produced rapid economic growth. This continues to make it attractive to the poorer countries, who feel that they cannot afford to slow their growth, or to impose higher costs, through experimenting with safer, slower, or more expensive paths to industrialization. Pollution is already severe in many industrializing cities in the poorer countries, and in their use of chemicals and pesticides agriculturally. One can only hope that the development of solar energy (especially appropriate since so many of the poorer countries are in areas of strong sunshine) or other clean sources can take place soon enough to rescue the Third World – and the rest of us, since we share a single ecosystem as we share a common human destiny.

The vital commonality of all people in these terms was brilliantly expressed by the British cartoonist David Low at the time of the Munich crisis in 1938, just before the outbreak of World War II. He showed a life-boat in a stormy sea, with a gaping hole in the bow. Clustered in that end were the countries of central Europe threatened (and soon to be over-whelmed) by Nazi conquest. In the drier stern were France and England, the countries which had through the Munich agreements given the Nazis in effect a free hand in central Europe, in exchange (they thought) for leaving them alone. England was saying to France: "Nasty leak up there; good thing it's not at our end of the boat." France and Britain learned the hard way that central Europe's agony was theirs. We need still to learn that the boat we all inhabit is the world itself, all its people, all their prob-lems, all their strivings for development, and the environment we jointly share.

15 RESOURCE USE, INDUSTRIALIZATION, AND THE FUTURE

"Industrialization" may be defined as the use of inanimate power and machines in large units or groups to produce uniform goods in much larger amounts and at a lower unit cost than was earlier possible without power-driven machines. "Industry" is a less precise word which may broadly cover almost any kind of productive enterprise, including agriculture. Ancient or medieval production may also be called industry. The change to what is defined above as industrial or industrialized came perhaps too slowly to be accurately called a revolution. It was spread over at least a century after about 1750, and it is clearly still going on as methods of production change rapidly and new machines, materials, and power sources come into use with what may justly be called a revolutionary speed and a revolutionary effect. Before the 1750s there was a much more gradual refinement and increase in production and techniques, although from the beginning of recorded human history to the eighteenth century there was very much less technological and mechanical change than has taken place in the short time since. These changes were especially accelerated after 1800, and their overwhelming effects on every aspect of human life were and are certainly revolutionary. The phrase "industrial revolution" is in common use, and as long as one understands the time measure involved and the nature of the change, it is a convenient label. Primarily, it reflects an enormous increase in the variety and quantity of resource use. Industrialization brings resources into being not only by discovering them but by making them usable at a bearable cost and with greater efficiency.

What makes it possible for potential resources to be used, or what brings resources into being? Why did industrialization take place on a revolutionary scale in much of Europe, North America, the Soviet Union, and Japan but has only recently begun in China and India, and has still to come in most of the rest of the world? If this question could be answered definitively, social science would have come a long way toward living up to its still inappropriate name. Unfortunately the answer can only be guessed at; too little is known and understood about it. Industrialization is still relatively new, and there is broadly only one case from which to generalize, that of Europe-North America, whose development was more joint than separate. Russia and Japan may perhaps provide important newer evidence of a somewhat different kind, as they have applied on a major scale in their own special situations commercial and industrial techniques originally adapted from the earlier Western model. The beginnings of major industrialization in China, India, Brazil, and elsewhere may constitute other examples. But it is hard to say whether any or all of these cases give enough data from which to form adequate conclusions. One can say for certain only that a great number of factors are involved in industrialization and economic development, and that each situation is to some degree unique. The factors include many things besides minerals, soils, climate, or other physical resources. Location, the size and nature of the market, transportation, capital, technology, social structure and cultural values, government policy, the nature, size, and skills of the labor force, managerial and entrepreneurial skills, the nature of economic organization, and popular attitude toward change or toward industrialization must be considered, plus other factors whose relevant importance is difficult to gauge. Presumably the critical factors must also be operative in the right proportions and at the right time. This is a fascinating and important problem to investigate, but it is outside the scope of the treatment here. We are concerned for the moment with what brings potential resources into being and makes them usable.

Resource use

Human knowledge and technical ability are clearly necessary to make any resource effective. The society must be aware of the resource as something offering support and must know how and must be technically able to use it. The additional factors which make a potential resource usable may be further simplified as the *cost of use* and the *price* offered or *demand* for it. Such an analysis is difficult to apply to resources like location or climate, except as the location of a resource affects the cost of its use, or to some of the things on which pre-industrial societies may depend, but it is particularly helpful for minerals, biotic resources, or land. Cost is a measure of human powers, and price or demand is a measure of human wants. Human powers and wants are seldom constant, and their interaction may invalidate resources as well as creating them. A resource comes into being, assuming someone knows how to use it, when the cost of its use is met by the price which the economy is willing to offer for it. If cost is greater than price, the material or potential advantage is not usable. The usability of any potential resource also depends not only on the cost of using it but on the cost of using some competitive substitute which may be cheap enough and may be available in great enough quantity to satisfy the demand entirely. Other iron ore or coal or power sources may be cheaper to use, or other land may be cheaper or more productive to irrigate.

Technology and changing resources

The cost factor is continually affected by technological change, which may make it possible to use economically deposits or land or situations which were not resources at all before. The climate of California became a much more valuable resource when low-cost transport links were developed with the American market as a whole, and when large-scale irrigation became possible. The same kind of analysis applies to the commercial usability of any locational advantages. Changing costs and technology also affect the usability of minerals. Iron and aluminum are found in nearly all soils and are two of the commonest elements in the earth's crust. In this diffuse form they are not now resources because it is so much cheaper to obtain iron and aluminum from the more concentrated deposits provided by nature in certain places. A rich iron ore will contain as much as 65 per cent pure metallic iron; generally, ores of less than 35 or 40 per cent have not until very recently been economically usable except in special situations. But suppose that someone developed a powerful magnetic extractor which could be passed over the ground like a vacuum cleaner picking up the iron. If the costs were low, this bit of iron in the soil might become a resource, provided costs for extracting and transporting iron from the existing ore deposits did not fall even lower.

The industrial revolution has depended on technological developments essentially like this example. The cost of obtaining and using raw materials and power has progressively been reduced so that a great variety of latent resources have become usable, including land which could not economically or productively be cultivated when technology was less advanced and costs of use were higher in proportion to the yield. The transformation of coal into power (rather than burning it only for heat as before), through the medium of the steam engine, one of the first modern technological developments, is an example of how industrialization has also changed and multiplied the usefulness of pre-existing resources. Extraction and transport costs for coal were also greatly reduced with the help of machinery and inanimate power, and coal became a much more important resource than it had been before. Aluminum has become a resource only recently. Technological improvements in extraction and refining have reduced the relative cost of aluminum by nearly three thousand times since 1860; with each cheapening it became usable for more and more purposes. Originally aluminium was used only for small, high-value goods like cooking utensils; now it may be used extensively for structural purposes or for bulky transportation equipment. Steel went through the same kind of process a hundred years ago, when the progressive lowering of costs made possible by improved technology made steel the cheapest suitable metal for most purposes. Aluminum has many superior physical properties, including resistance to corrosion and great strength in proportion to its weight, but it cannot displace steel for many of the uses where it would physically be suitable as long as it is more expensive. Copper also has many desirable properties and would be preferable to both steel and aluminum in many cases, but the higher cost of copper at present confines it to uses where its special properties are particularly needed, such as for conducting electricity. Many metals are interchangeable for a variety of uses; relative costs mainly determine which metal is used. Relative costs may change, and may thus affect resource use and resource appraisal. Copper and aluminum are particularly close competitors for a number of uses, so that slight changes in relative costs can have substantial effects on the consumption of both. One development which helped to make steel cheap was the perfection of low-cost techniques to use lower-grade ores or ores containing impurities (especially phosphorus) which were unusable before. This also had the effect of greatly increasing the total reserves of iron ore and fundamentally changing the resources inventory. France, for example, found itself, as a result of this discovery, well endowed with iron ore, whereas earlier usable French ores were limited because of high phosphorus content.

As a resource continues to be used, the criteria applied to it may change. Copper, for example, has been mined in the United States long enough to exhaust the richest deposits. Fifty years ago material containing less than 4 per cent pure copper was discarded as worthless; now material containing only 1 per cent copper is considered usable ore, and in some cases it pays to work over the discard piles of fifty years ago. Technological change is largely responsible for this, since it has so greatly reduced the costs of extraction and refining. But in the Congo Basin, where rich copper ores have only recently begun to be mined, 4 per cent is still a basic acceptable minimum. The original deposits of rich iron ore in the Mesabi Range near the western end of Lake Superior are approaching exhaustion. They have been the major source of iron ore for American manufacturing not only because of their high quality but because of their fortunate location close to the water transport network of the Great Lakes, even though they are not very close to the major markets in real distance. As these ores approach exhaustion, new steel plants have been built on the east coast, where ores can be brought in from overseas by cheap water transport. But there are also extensive deposits in Minnesota of a rock called *taconite* which is 25–35 per cent iron. Until recently it has been too expensive to get the iron out of taconite, but a process of extraction and refining or concentration called *pelletization* has now been developed which can produce iron from taconite at a cost competitive with sources of iron. Taconite has become an ore and has thus become a resource.

Changes in demand

Technological change also affects price or demand, since it changes the market's ability to use things and may create or remove needs for certain goods. There is no longer a demand for whalebone for corset stays, and the demand for buggy whips, goose grease, woad, or wolfbane has greatly decreased, not only because cultural preferences have changed but also because technological change has brought about substitute demands or has developed alternative goods or sources which are cheaper or more effective. Totally new demands may also arise. A hundred years ago there was little demand for rubber except as erasers – this was the origin of the name *rubber*. The development of the automobile in particular has created a tremendous market appetite for rubber, as industrialization has created a great variety of new demands for other new resources. A hundred years ago petroleum had almost no uses except as medicine; electric power was just growing out of the toy stage and had few uses. Technological change reduced the cost of obtaining and transporting the new resources, but it was also responsible for bringing about the demand for them in the first place, or, in other words, for bringing the resources into being. As demand increased, efforts were further concentrated on increasing production and lowering costs. It was the interaction of increasing and expanding demand and falling costs which created new resources, as in other cases declining demand removed items like whale-bone or wolfbane from the resource list.

When real or artificial shortages arise, as for example in wartime, price rises or increases in demand may also make resources out of land, minerals, commodities, or strategic locations which were not resources before, or whose cost kept them from being usable when the price or demand was lower. Mines abandoned because the richest deposits were exhausted may be reopened if ore prices rise or if the minimum acceptable standards for ore fall. Abandoned or marginal land may be cultivated. Normally uneconomic substitutes of various kinds may be pressed into use. In normal times areas which lack coal may develop hydro power sites which would not be developed in coal-rich areas. Areas which are poorly endowed with mineral resources in general may use deposits or materials which would not be considered resources elsewhere. Demand may thus create or eliminate resources by affecting the resource appraisal. But technological change tends primarily to reduce costs and to widen demand rather than to raise prices, and it is in this way that most new resources come into use. The rising cost of fossil fuels is bringing solar and other clean energy sources rapidly closer to mass use. Many more resources are now used than before the industrial revolution, including many which had little value or were even unknown a hundred years ago. This is usually a cumulative process, for very few resources in the general sense, as opposed to specific deposits or individual sources, pass out of use entirely. We still make use of nearly all of the things used in the distant and more recent past, although we may attach much less relative importance to certain things. Each year new resources are added to the list, and it is difficult to predict what may be included in the future. The rate of technological change is still accelerating.

What people depend on

The above discussion of land, minerals, or other commodities should not draw attention away from the broader and valid concept of resources as it was defined earlier: things on which we depend. Climate or soil may be a more important resource in this sense than coal, as may location or animals or fresh water. Sea water may become a major resource if the variety of minerals which it contains can be extracted from it cheaply enough to meet or beat the cost of present sources. Resources like climate or location are, however, difficult to evaluate completely in precise cost terms. Technological change affects the usability of climate or location as it affects other resources; they become more or less usable in accordance with changes in human powers and wants, as indicated by the examples of California's climate or Japan's location. Two of the leading manufacturing industries of California, airplanes and films, depended in part, especially in their early growth, on the resource of the local climate, as Japan's economic development is dependent in part on the resource of her location. Such resources are directly relevant to industrialization, but since their effects are less easily reduced to detailed measures, it may be more helpful to analyze selected samples of resources which can be so treated and which will illustrate the general problem of resource usability.

Resources and the location of manufacturing

Of the many thousand raw materials on which an industrial economy is dependent, a few are of paramount importance, especially as they help to determine where industrialization may occur. These few are important because modern manufacturing uses very large amounts of them and therefore cannot normally afford to haul them from great distances. The mineral alloys (for example, chrome, vanadium, nickel, or tungsten) which are necessary to make steel hard or resistant to heat or corrosion are just as necessary as iron ore for the making of that steel. But such relatively small amounts of these alloys are needed for every ton of steel that they can economically be brought from the other side of the world if necessary. Some things which are used on a large scale occur or can be produced only in parts of the world where there are no other advantages for manufacturing; they therefore do move considerable distances to reach the established manufacturing centers. High-value tropical crops like rubber, chocolate, or vegetable oils do this, and of course petroleum in far larger quantities. Mineral deposits occur in widely scattered locations, and only by chance do they coincide with favorable climate, land forms, population centers, or other factors conducive to manufacturing. When they do not, they may not be usable, if they are bulky or heavy in proportion to their value.

There are only a limited number of major heavy manufacturing areas in the world (see Map 26): north Europe, the Ukraine-Urals area in the Soviet Union, northeast United States, central and western Japan, north

Map 26 Major world manufacturing areas. *Data are for the late 1970s, slightly different years for different countries. Distinctions are based on total value of industrial output from power-driven plants employing more than ten workers.*

China-Manchuria, southeastern Brazil, and the Damodar region of India, but there are plenty of mineral deposits distributed around in other places. Heavy manufacturing, in which raw materials and transport costs are a large proportion of total production costs, must concentrate where assembly costs for its raw materials and distribution costs for its finished products to market are at their lowest. It must be particularly sensitive to its economic distances from the raw materials of which it uses the largest amounts. If these economic distances are small, and if distances to market are also small, heavy manufacturing can ignore distances for other materials which it needs in lesser amounts. Large deposits of copper, aluminum, manganese, antimony, tungsten, or nickel are unlikely by themselves to attract much manufacturing locally, aside from the immediate processing of the ores, because they are used in relatively small total amounts in manufacturing as a whole. If an ore contains only a small proportion of pure metal, it may not pay to ship long distances what is mostly waste material. Such ore is usually somewhat refined on the spot. Copper ore, for example, is often smelted, or concentrated into *blister* copper, at the mines; what is sent to market or to copper-using industries from smelters is about 99 per cent pure copper. But heavy or general manufacturing would find little advantage at a copper or tungsten mine if the mine were far from sources of fuel and power, from other bulky raw materials, from labor, and from the market.

Heavy manufacturing and its role

Heavy manufacturing, which includes most importantly iron and steel, construction materials, transport equipment, chemicals and refining, and heavy machinery, is fundamental to industrialization because it provides the essential means of producing and exchanging more goods by power-driven machinery. Without these basic essentials, or *capital goods* (sometimes called *producers' good*), it is difficult or impossible to produce or exchange *consumer goods* like textiles, furniture, or food for wider markets, except in small quantities or at a relatively high cost. Thus all industrialization in the modern sense depends on the heavy industries. The supply and cheapness of fresh and canned foods, ready-made clothing, houses, or beds depend upon steel, machinery, and transport equipment. Consumer-goods industries, most of which are mainly market-oriented, since raw materials are for them a relatively small proportion of total production costs, are often attracted to centers of heavy manufacturing and their attendant markets, power, services, capital, labor, and transport facilities. Consumer goods are made nearly everywhere in some form and to some degree and have been for thousands of years. But they are made cheaply and in large quantities only where they can depend on heavy industries to supply them with the means of mechanical production and exchange.

The big three: coal, iron, and oil

Coal, iron ore, and petroleum form a triad of heavily used resources to which most manufacturing must be especially sensitive. Fresh water

COAL, OIL AND IRON ORE 1977

Iron Ore Producing Districts
(Millions of tons per year)
  o 1 - 15
  O over 15

Coal Fields
  Major Bituminous Field
  Minor Bituminous Field
  Anthracite
  Lignite

Oil Fields(simplified)
  Major Producing Region
  ● Major Field
  ○ Minor Field

Map 27 World coal, iron, and oil. *Data are for the late 1970s, slightly different years for different countries.*

might well be included, since it is used in even larger amounts by most manufacturing, but fresh water is more widely distributed and for the most part influences the location of manufacturing on a smaller or more local scale.

SCALE

GOODE HOMOLOSINE EQUAL-AREA PROJECTION

(1) *Coal*. Of the big three, coal was long the most important, because it was the cheapest available material for a great variety of uses and was therefore consumed in huge quantities. As pointed out earlier, coal is often important enough in this sense to draw to it whatever other raw materials are needed and thus to minimize total transport or assembly costs, if the factors of location, market, technology, capital, labor, and transport are favorable enough in the area where the coal occurs. There are very few large industrial areas which are not near or accessible to coal, and most of the largest ones are very close to coal, at least in economic distance. Coal is used not only in the metallurgical industries as coke (see note 1 on page 189), but as the largest single source of electric power world-wide (from steam plants, where power is generated by steam-driven turbines), as a leading raw material in a number of chemical industries, where it is especially important for production of synthetic substances such as dyes and plastics, as a fuel for heating, and for many other purposes. As petroleum, hydroelectricity, natural gas, and fission-able materials have become increasingly important sources of power, coal has lost some of its former proportional dominance. For example, coal is no longer the chief fuel for domestic heating or for railways or ships. But for many years coal will remain the cheapest material for a variety of manufacturing needs, and it continues to be the most used industrial raw material, after fresh water.

It is possible that coal's relative importance may increase in future, at least for a time, since world petroleum supplies will almost certainly be exhausted long before world coal deposits. Oil can be obtained from oil shales, which are relatively plentiful, but at what may in future be a higher cost than oil from coal. Petroleum or its products can be made from coal, as is done already on a small scale in certain areas where petroleum is scarce. It will certainly increase in the United States, however short-sightedly in view of the destructive impact on air quality as indicated in Chapter 14. Controls to limit pollution from coal are technologically easy enough, but there is resistance to their cost – as to adequate automotive emission controls. The coal-rich areas will probably continue to enjoy major economic advantages, as they have since the eighteenth century. Much of the enormous and rapid industrial development of the United States was due to the fact that it is "the land of cheap coal," with about half of the world's total workable coal reserves, plus large shares of the world's petroleum, iron ore, and other valuable minerals. Extraction costs for coal are also relatively lower in the United States than in most other coal-mining areas, due to the favorable nature of most of the deposits. Huge coal deposits of somewhat lower grade but with very low extraction costs, close to the surface and accessible by strip mining (removing shallow surface layers) are still largely untapped in several western and mountain states. The rising price of all fuels now makes these reserves more attractive economically, despite their distance from most major markets. But the United States is rapidly consuming its minerals and is already a large importer. With the fortunate exception of coal, we may shortly be a net importer of most of the heavily used minerals. American use rates are many times the use rates in the rest of the world and are still rising.

(2) *Iron Ore*. Since iron ore is used in somewhat smaller total amounts than coal, it therefore is usually transported to places where coal is cheap. But manufacturing uses enough iron ore to have its location profoundly affected by ore availability, as illustrated by the trend of migration of the iron and steel manufacturing in the United States toward the east coast in response to changed supplies of iron ore. Technical economies in the use of fuel have decreased the proportional amount of coal necessary for iron and steel making (and for other manufacturing processes also) so that slightly more iron ore than coal by weight is now used in making a ton of pig iron. But coal has many other uses, and steel manufacturing also makes large use of scrap metal as a raw material, mixing it with pig iron to make steel and to some extent replacing iron ore. Scrap is usually cheapest and most plentiful in the large market centers, which are often associated with coal, import coal, and consume large amounts of metal products. Thus coal's locational dominance tends to remain, as well as its association with the dense populations necessary for manufacturing, supported by manufacturing, and constituting large markets; this association is clear, for example, in Europe, the oldest and still the largest industrial area, as shown in Map 28.

Markets, the transport facilities which serve them, and their pools of labor, capital, special services (such as engineering, banking, or insurance), and consumers are a major attraction for nearly every kind of manufacturing, wherever transport is cheap and adequate enough to bring the necessary raw materials to the market at a bearable cost. Coal is closely associated with markets and dense populations in so many cases because coal is too bulky and low in value to be transported far in the huge quantities needed, and the markets and populations have thus frequently grown up close to the coal. As transport facilities improve and transport costs for bulky raw materials decrease, the market becomes an even more desirable or profitable location for manufacturing. This process has continued throughout the industrial revolution so far, as the sites of raw material and power sources have increasingly lost their locational attraction and the market has gained. Since so many of the markets grew up in close association with coal, and since coal is still used in such large amounts, this development has not affected coal as much as it has affected other raw materials.

Where water transport is available, iron ore may move considerable distances to market centers where coal, scrap metal, other necessary raw materials and power, labor, capital, and transport facilities are present. Specialized carriers also help to make the movement of iron ore relatively cheap despite its bulk and weight.

(3) *Oil*. Petroleum and natural gas are much more foot-loose than coal or

Map 28 Coal, iron, and population in Europe. *Data are for the late 1970s, slightly different years for different countries.*

iron ore, even though they have also come to be used in huge quantities. They are much more cheaply transportable, by pipeline overland (almost as cheap as ocean transport) and by tanker at sea. They can be loaded and unloaded quickly, entirely mechanically, and thus cheaply; special carriers built to haul nothing but petroleum and loaded to capacity with no waste space can offer very low haulage rates. The relatively low transport cost for petroleum is fortunate, since so much of the world's petroleum lies in the Middle East, the East Indies, Mexico, and Venezuela, far from major market centers.

Petroleum alone does not create industrialization where the deposits occur. Oil can too easily be moved to places which have greater total advantages for manufacturing. The oil of Iraq or Saudi Arabia, for example, is not necessarily a stimulus to local industrialization except as it can provide capital which may be invested in manufacturing; other requisites for industrialization, such as other raw materials, technical skills, adequate political leadership, market, and skilled labor are largely lacking, and oil alone cannot attract or create them.

What makes a mineral deposit usable

Mineral resources are only part of the resource concept. They are a distinctive part because, unlike climate or water power, they can be exhausted by use and are not renewable; this imposes more rigid economic conditions on their exploitation. Mineral ores and fuels are the most important members of this group; the conditions of their exploitation can be taken as typical of other minerals, and in a broader sense of all resources. What helps to make a given mineral deposit usable as a resource? The factors include the following:

Physical factors

(1) *The size of the individual seam, bed, or pool.* Particularly with mining, this will be reflected in costs. If the coal seam is thin, or bent and broken, it will cost much more to get the coal out than if it occurs in a thick regular layer. If the seam is too small to permit mechanical mining, costs rise steeply. This has happened in many of the older coal mines in Great Britain and Japan, where the thickest seams have been worked out. The United States is still skimming the cream of its coal deposits, and costs are correspondingly lower. The same considerations apply to ores and to petroleum.

(2) *The depth below the surface at which the material lies.* The relation of this to the cost of extraction is obvious, though it is often ignored by "resource" surveys. What lies on top (the overburden, as it is called) must either be removed or deep shafts sunk, and thus heavy costs are undergone before there is any yield. Inadequate size or excessive depth are largely responsible for the fact that more ores and coal must be left in the ground than can economically be brought to the surface, even where a mine has been dug. If present experiments with burning coal in place underground can produce an economic source of power or by-products, the resource status of nearly all coal deposits would be profoundly affected.

(3) *The quality of the material.* Standards for this may change, as discussed above, but in any given situation the grade of the ore, coal, or petroleum will help to determine whether the costs of extraction can be repaid. Naturally, a high-grade deposit will make more expensive extraction worthwhile.

(4) *The size of the total workable deposit.* This also frequently leads resource surveys astray. For modern industrial operations, a deposit must be large enough to enjoy the economies of scale. To act as a localizing factor, it must also be big enough to support the industrial operation for a long time. A steel plant, for example, cannot afford to pick its location on the basis of an ore deposit which it would exhaust in two years. In practice, this factor rules out a great many deposits for which the other factors are favorable or which were used earlier when the scale of manufacturing was smaller. Small bodies of iron ore in New Jersey and Massachusetts, for example, which once supported a flourishing industry in the eighteenth century, using charcoal as fuel, are now of only academic interest, mainly because of their limited size. This is typical of similar small bodies of ore in a great many places elsewhere, many of them once smelted with local charcoal but now obsolete even though they may be of high grade and have low extraction costs. Improvements in transportation, widening of the market, the change to dependence on coal rather than on charcoal, and the greatly increased size of manufacturing units with attendant lower costs of production have resulted in a concentration of production at a few sites where very large supplies of ore and coal are cheaply available. State or national totals of mineral sources may thus mean very little.

(5) *The distance from the market or from the manufacturing site.* This will mean primarily economic distance, reflecting topography, other aspects of the intervening space, and available transport facilities as well as physical distance.

Economic factors

(1) *The size and nature of the market, or the effective demand.* The market must be willing and able to pay for raw materials and to use them profitably. Most pre-industrial economies cannot do either on a large scale; hence their coal or ores may not be resources for them.

(2) *The availability and cost of capital.* This is primarily a reflection of per capita incomes or of the size of the economic margin. Plentiful capital at low interest rates is essential for industrialization and for the use of industrial resources. The economy must be able to save, or to set aside surplus goods and postpone consumption, in order to meet the heavy costs of building a factory or a railroad; these capital goods, or producer's goods, cannot be eaten or worn but are used to make and exchange goods which can be consumed. The increases in the production and exchange of consumer goods which the production of capital goods makes possible do not take place immediately, and there may be a wait of several years before the results are available to the consumer. Without ample surpluses or capital for investment, this postponement of consumption and its later

rewards is not possible; industrialization cannot take place, coal mines cannot be dug, and latent resources cannot be used. If investment capital is expensive, as it nearly always is wherever it is scarce, resource use and industrialization are retarded.

(3) *The transport media.* Without low-cost volume carriers, bulky goods cannot economically be moved, and many materials cannot be used.

(4) *The technological means available.* These have already been discussed.

All of the above factors help to determine in any given case whether a particular material is usable and whether cost and price or demand are in appropriate balance. The weight of these factors differs at different times or in different places. Japan, with a domestic coal shortage and few rich deposits left to draw on, mines coal which would be ignored in the United States. But China, with much good-quality coal, has until recently used relatively little of it, either because its location was wrong with respect to the market, or because the market and the supply of capital, transport, and technology were unfavorable. The Japanese market calls forth what in other situations would be impossibly high-cost coal because there are no cheaper alternatives available, and makes proportionately large use of oil to generate electric power, since oil can be shipped more cheaply than coal. Domestic resource inventories also help to account for the development of hydroelectric power in coal-poor Italy and France as well as in Japan. Thus resource appraisals differ with differing economic as well as technical and cultural conditions. A mineral deposit can be "exhausted," or re-worked, several times as both technological and economic conditions change.

In Chapter 9 and Chapter 12 the same kind of analysis was applied to timber as a resource. There the economic distance to market is particularly important because timber is so bulky and heavy in proportion to its value and therefore cannot bear expensive transport hauls. Problems of mining are roughly paralleled by the problems of getting trees from steep slopes in remote mountains or out of the barrier country of the rain forest, where mobility is limited and where there are local physical obstacles contributing to high extraction costs. Thus many forest stands are not resources because they are not economically usable; cost and price are not in appropriate balance. Chapter 10 analyzed soils as resources in the same way. The set of factors outlined above helps to determine the usability of all mineral deposits, including the many non-metallic ones like salt, building stone, sand, gypsum, nitrates, or sulphur and also affects potential resources which are less concrete, like climate or location.

Resource crisis or a brave new world?

There is a continuing public and private debate between those who are alarmed at the rapid depletion of mineral, water, biotic, and soil resources and those who argue that the equally rapid progress of technology will create new resources by finding new sources of power and materials or by decreasing the cost of using the present ones. One side stresses the undeniable fact that world population is still increasing and may increase more rapidly as industrialization spreads (see Figure 24). Good agricultural land is limited, and possibilities for expanding profitable cultivation may not be great enough to compensate for the larger number of mouths to feed. While only about 10 per cent of the earth's land surface is now cultivated, the remaining 90 per cent (including the deserts and ice caps) is neglected for good reasons. Any major extension of cultivation is likely to involve the use of marginal land. This not only presents cost problems but means that, whatever the cost, such land is not likely to be as productive as the existing cultivated land. Rising market demand and prices may make it profitable to cultivate new land, but the usable supply is restricted, and the returns diminish as cultivation is extended onto less and less good land. At the same time, much cultivated land is being destroyed or abandoned as a result of urban-industrial spread, erosion, or soil poisoning/waterlogging from over-irrigation.

Mineral ores and fuels are rapidly being used up. The world consumed more minerals in the first half of the twentieth century than in all previous history; the rate of use is still rising and will continue to rise, particularly as more of the world becomes industrialized and as population continues to grow. Half of all the coal ever mined has been dug since 1940, half of all oil since 1965! These resources, which took 100,000 years or more to form, will have been largely depleted in about a hundred years of large-scale industrialization. It is argued that technology cannot increase resources or production in any sphere fast enough to keep pace with population and consumption increases, especially when it is confronted with finite or disappearing supplies of land and raw material.

This is essentially the same dilemma which Malthus described for pre-industrial Europe and which he mistakenly assumed would continue in the nineteenth century. The "Neo-Malthusians" argue that because the West escaped temporarily from the unfortunate consequences of that dilemma is no guarantee that the West, and more especially the rest of the world, can continue to escape in future as world population growth gets into its stride and available resources are depleted. Some estimates suggest that tin, lead, and zinc, in terms of their currently classified ores, may be largely exhausted by the end of the present century at current or expected rates of use; so may commercial timber by present standards, and iron ore may very shortly follow. Deposits of coking coal and petroleum which are now considered economically usable may all have been consumed in another century.

The reply is the truism that resource appraisals are not constant; they change as people change. At least so far, as rich ores, deposits, soils, or forests have been exhausted or taken up, stronger demand, higher prices, or lower-cost and more efficient techniques of use, or a combination of all of these things, have often made it possible to use leaner ores, lower-grade and more remote deposits or forests, oil shales, or poorer land without necessarily decreasing average yields or per capita consumption. Better recovery techniques may also make it possible to get more of the oil, coal, or ore out of a given deposit; at present, especially with coal and oil, more is left buried than can be taken out, although proportionately more and more oil is being recovered by pumping and other methods. New deposits are also almost certain to be discovered. Even in relatively well-surveyed areas like the United States, new mineral concentrations are continually being found. Admittedly, this cannot go on indefinitely. Total world coal reserves may last for another two thousand years, but for coking coal and most other minerals, especially the metals and petroleum, the end may already be in sight in terms of present techniques. Equally alarming is the galloping rate at which we are destroying vital resources of air, water, and soil through pollution.

The crux of the matter is that techniques change and are now changing particularly rapidly; this means in effect a change in resources. As awareness of the frightening consequences of pollution and chemical waste disposal grows, we may also hope that this suicidal damage to the environment can be checked. We have the technology to prevent pollution, and to dispose of wastes cleanly. We need to apply it, and to accept the relatively small added costs involved – for our own survival. Yields of all resources are more likely to increase than to decrease, judging from past experience, although there can be no assurance that yields can increase as much or as fast as population or consumption, if present growth rates remain unchecked, or that pollution can be halted before the environment is irreparably damaged. Accessible sources of charcoal, for example, were largely exhausted in England and elsewhere by the eighteenth century; the metallurgical industries of that time were rescued from what appeared to be a major and crippling resource crisis by the development of coal and coke as metallurgical fuels. The same would apply to the shortage of sizable oaks for ships of the line and the rescue of the British navy by the development of iron and steel ships.

If and when the bottom of the barrel is reached in terms of presently known and appraised materials, as it may well be in the not very distant future, the hope is that technology will have progressed enough to extract at a bearable cost whatever minerals are needed from almost any rocks or soil or from sea water. Fuels and power may be derived from the sun. The problem of pollution control is political, economic, and social rather than technical. Food and plant cellulose may be produced by chemical synthesis, by artificially manipulated and accelerated growth of certain prolific plants like algae and yeast, or by planned cultivation of the oceans and the continental shelves, as well as by a more productive use of the land and of fish, animal, insect, and bird resources. Some enthusiasts argue that in another two or three centuries the world might support twenty or thirty billion (US) people at a much higher average income or consumption level than the present world population of about four and a half billion enjoys. But there remains the fundamental and disturbing question of whether the production of food and other goods can in fact be increased fast enough and at a bearable cost to keep pace with galloping population increases, especially in the so-called underdeveloped world. The two sides of this controversy are amusingly pictured in the following impromptu lines, written by Professor Kenneth Boulding while he was attending a conference on "Man's Role in Changing the Face of the Earth:"

A conservationist's lament

The world is finite, resources scarce.
Things are bad and will be worse.
Coal is burned and gas exploded,
Forests cut and soil eroded.
Wells are dry, and air's polluted,
Dust is blowing, trees uprooted.
Oil is going, ores depleted,
Drains receive what is excreted.
Land is sinking, seas are rising –
Man is far too enterprising.
Fire will rage with man to fan it
Soon we'll have a plundered planet.
People breed like fertile rabbits
People have disgusting habits.

    Moral:
        The evolutionary plan
        Went astray by evolving man.

The technologist's reply

Man's potential is quite terrific;
You can't go back to the Neolithic.
The cream is there for us to skim it –
Knowledge is power, the sky's the limit.
Every mouth has hands to feed it,
Food is found where people need it.
All we need is found in granite
Once we have the men to plan it.
Yeast and algae give us meat,
Soil is almost obsolete.
Man can grow to pastures greener
Till all the earth is Pasadena.

    Moral:
        Man's a nuisance, man's a crackpot,
        But only man can hit the jackpot.

There is something to be said for the criticism that conservation some-times appears to put a higher value on non-human materials and their preservation than on people whom they serve, and that human ability to create resources is often underestimated. It is common sense to conserve renewable resources like soil, water, or the biotic world in general by wise use which allows or encourages replacement of what is used, in cases where the return equals the costs of conservation. It is also intelligent and may soon be essential to use non-renewable resources as carefully and as efficiently as possible, although enough may not be known in any particular situation to decide whether it is best to use or to conserve. But the essence of resources is use, not meaningless restriction. Sound and positive conservation means only that misuse is avoided, and that the environmental base is preserved or permitted to renew itself.

Future resources

No one can be certain how far or how rapidly technological change or other factors can alter resource use, although the record of the past hundred years in this respect suggests that the rate and degree of change have consistently been underestimated. One can be sure only that the meaning and use of resources are highly changeable and that the future cannot accurately be measured in terms of the present. We may well be on the threshold of a vast new development in resource use which will add new resources to the list and change the meaning of many present ones. But because a material or advantage is technically usable does not necessarily mean that it is economically so, if total costs of use are too high. It is also generally unwise to assume that when new resources come into being the present locational pattern of economic activity will be greatly distorted. It might be in certain cases, as it clearly was by the advent of coal as the single most important mineral resource of the modern world. But because the Congo Basin has much uranium, for example, does not mean that it will become a great manufacturing or population center on the basis of nuclear power. As with resources in the past and present, assembly and distribution costs will continue to operate. The broad, varied bases of settlement, manufacturing, transport, markets, and capital will continue to be more important than specific mineral deposits in themselves, even or especially if all rocks or all soils should become usable sources of minerals. General technological change is more likely to benefit established centers than presently underdeveloped areas. Another decade or two might see some change in this relationship. Europe's experience with population growth may not be applicable to Asia, for example, but it will be difficult for the presently underdeveloped areas to reach the point where their per capita production increases as rapidly or more rapidly than Europe's at present. The development of cost-efficient solar power could be an enormous boost to the developing areas of the poor world.

Additions to the list of usable resources, including many of a less dramatic kind than nuclear or solar energy, may continue to increase the standard of living, and one may hope also the common ground between nations. In a world of relative plenty where goods are better distributed and resources more widely used, there may be less occasion for discord. Such possibilities are of transcendent importance for all people. But resources are not exclusively physical phenomena so much as they are reflections and creations of the human factor. This is apparent in the tremendous regional diversity of the world and in the particular patterns of resource use which have grown up in its different parts. The materials or advantages on which civilizations and economies depend vary enormously. Resource appraisals vary as places, people, and societies vary. One of the greatest challenges to modern civilization is whether it can increase and spread resource use widely enough so that, by varying local combinations of resources and wants, economic abundance is available to all.

16 CITIES AND MANUFACTURING: SPATIAL CONCENTRATION

We live in an increasingly urban world. Not yet most people, but the great majority of people in all of the so-called "developed" countries now live in cities, and the proportion everywhere is rising. Figure 25 shows how rural and urban dwellers are divided within each major world region, and makes it clear that North Americans are overwhelmingly city folk. The proportion of a total population living in cities, or the *level of urbanization*, varies widely, as Figure 25 suggests, but is closely correlated with the overall level of economic development. This is usually defined in terms of the share of manufacturing in the total economy, the degree to which the economy is commercialized, the prominence of *service* industries (government, education, medicine, trade, banking, etc.), the level of per capita incomes, life expectancy, per capita use of energy, and other criteria by which development is measured. Modern cities are primarily the product of industrialization, as well as its major bases. They are also

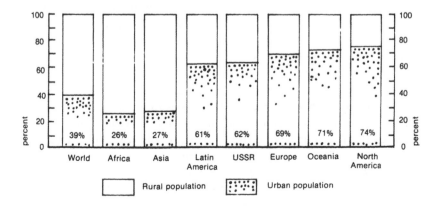

Figure 25 **Urban-rural divisions by continents**

the chief centers of the consequences and the benefits of industrialization, good and bad.

Cities have existed ever since the dawn of human civilization, and indeed our word "city" is derived from the same root as our word "civilization." Pre-industrial cities were centers of trade, administration, and artisan production, and were also important as cultural or religious centers. But the nature of cities changed fundamentally with the coming of industrialization, beginning in England in the late eighteenth century and spreading from there to Europe, North America, Russia, Japan, Australia, parts of Latin America and South Africa, and now increasingly in most of the rest of the world. Machine manufacturing is most conveniently and cheaply done in large units (factories) in concentrated centers: cities. As industrialization has spread, more and more people have been employed in manufacturing, and hence in cities, where there is additional employment in the service industries which higher incomes make possible or more complex development makes necessary.

Although the early stages of industrialization have been associated everywhere with poor working conditions, poor housing, and urban slums, and although industrialization also generates dangerous pollution as well as social tensions and human alienation, it is the essential means for the escape from poverty, recurrent famine, disease, and peasant toil which were the lot of most people everywhere until quite recently. In time, the new productive powers of industrialization result in an improvement in the material standard of living for most people, even though this "trickle-down" process may seem long delayed. Many traditional values are lost in grimy industrial cities, and many new problems spawned, but new wealth is also created, and its rewards are attractive to people everywhere.

Cities as progress?

Cities are also centers of institutional, intellectual, artistic, and technical change and achievement. This has always been their nature, and is part of the reason why they have grown. Cities are the center stage of any society. But they also mirror, perhaps magnify, society's problems as well as its achievements. In the United States, it is in our cities that we find the worst of our problems: crime, slums, violence, pollution, de-humanization, and also our greatest cultural, economic, and technological success. In most industrialized and highly urbanized societies many people have reacted critically to the undeniable problems posed by cities, or have even regarded them as a backward step, as if we would somehow be better off without them. There is no denying that cities have created new ugliness, new squalor, new versions of poverty and injustice, new and destructive assaults on the environment, and new de-humanization of life, or that in many ways rural life seems more attractive. But it is also cities and the things which happen in cities which have made possible most of the bases of modern life which we take for granted: electricity, hospitals, pharmaceutical drugs to combat illness, fertilizers, machines, and research to make agriculture enormously more productive, other machines which serve people or make their lives both easier and more rewarding, central heating and hot water, quick safe travel, and so on, not to mention universities, museums, orchestras, libraries, and other urban contributions to cultural life. Thomas Jefferson's vision of a rural-based America, or more recently Mao Tse-tung's vision of a rural-based China, may still seem attractive, but there are surely very few people who would choose the full consequences of a rural life without urban-based industrial supports: famine, disease, and mass poverty, at least by comparison with modern industrial living standards. Unfortunately, economic development comes in a package and it is really not possible to pick and choose; one must take the bad with the good, chemical industries as the price for anesthetics, and so on. It makes better sense to concentrate attention on trying to lessen the problems caused by cities, especially environmental pollution and social injustice, than to yearn for a return to a traditional past which can never be reclaimed, and which even if it could would no longer satisfy people used to what the modern industrial world can offer.

Cities have multiplied because of the advantages of *concentration* for manufacturing and the service industries, and because industrialization has also revolutionized the means of transport. Urban products and services can be distributed and raw materials collected widely at relatively low cost; the largest and most concentrated production units tend to be the cheapest and the most efficient. Efforts to distribute industrial units more widely, or to break them up into smaller units, have found in most cases that this increases costs. Some of the developing countries, notably China and India, have tried to do this, in a reaction against Western colonial models and in an affirmation of their own indigenous rural traditions. They have tried to explore new paths to industrialization and modernization which can avoid the many problems of urban concentration, stressing small rural industrial units and also the dispersal of larger complexes more widely around each country. This second objective has had some success, but it has understandably had the effect merely of creating new or much larger industrial cities in provincial areas; at the same time, the previously existing cities have continued to grow, in general the largest most of all, for the good reason that they remain the most cost-efficient centers of production and services, which is what made them grow in the first place. Small-scale rural industry has with few exceptions been less successful; it remains as a useful symbol, and as a means of absorbing underemployed rural labor to some local benefit, but in most cases it cannot compete in costs, quality, or efficiency with larger urban-based units. While this is indeed regrettable from many points of view, the developing countries urgently need to follow the quickest and the cheapest path to industrialization, and on a mass basis in order to care for their large and rapidly growing populations. They cannot afford to divert scarce capital or other resources, including skilled labor, into enterprises which, however desirable on other grounds, are slower or more expensive or which cannot produce in adequate volume or quantity. As transport facilities improve, the advantages of concentration become even stronger, and urbanization becomes even more dominant, as in the industrialized or developed countries.

Cities in the developing world now exhibit the same set of problems as in the industrialized world a century or so ago when industrialization was still in its early stages. There is fearful overcrowding, massive slum areas and generally inadequate housing, inadequate water and power supply, inadequate health and education facilities, and widespread unemployment as people are drawn in from the countryside in the hope of upward mobility but in larger numbers than the urban economies can quickly absorb. Calcutta, Mexico City, or Cairo seem to many Western observers like truly dreadful places, as of course in many ways they are. But they are also the hope of the future, in the sense that they are the chief *growing points* of economies still in the early stages of industrialization and economic development. In addition, we have enough information about them, and about the history of cities in the industrializing West in the nineteenth century, to know that in fact living conditions, including life expectancy, for most people in even the worst of contemporary cities in the developing world are almost certainly *better* than for most people in early nineteenth-century London or Manchester or in later nineteenth-century New York or Chicago. People forget quickly, and perhaps also forget that the process of economic development is hard, slow, and produces much misery for many people in the process of creating the means to a better life.

The history of cities

The first cities, probably in the hilly semi-arid area around the northern edge of the Tigris-Euphrates valley, in what is now southern Turkey and northern Syria and Iraq, evolved very slowly from agricultural villages to larger permanent settlements which began to be based in part on something more than simple agriculture: trade and the processing of agricultural goods. The gradual shift from agriculture or hunting *(primary)* to processing or manufacturing *(secondary)* and trade *(tertiary)* seems to have begun perhaps as early as about 9000 BC in these areas, and probably to have begun independently not very long afterward in Egypt, the edges of the Indus Valley (now in Pakistan), and perhaps in parts of Iran and elsewhere. Agriculture slowly became productive enough so that a small part of the population could be maintained in other occupations: making and exchanging things. As this *division of labor* increased, and as more non-farmers could be supported, the basis for cities as opposed to farm villages began to emerge. Cities are centers of non-agricultural production, exchange, and services, even though these functions may be performed on behalf of a rural hinterland, as explained in Chapter 3.

Soon after 4000 BC, larger and more specialized urban centers began to appear in the Tigris-Euphrates lowland. Once the basic techniques of agriculture had been mastered and populations increased, it became possible to work the far more productive alluvial soils in the river valleys, to protect them against flood damage, to use the rivers as sources of irrigation water, and to collect, transport, store, and exchange agricultural surpluses, for the support of new urban populations and as a basis of trade. By about the same period cities were also emerging in the lower Nile valley, and some centuries later in the Indus valley. All these areas had in common what all cities need: a productive tributary area which also supplied surplus food to maintain urban non-farm workers, an open and largely treeless plain across which movement was easy, and a navigable river to make bulk transport still easier and cheaper. Cities depend on *access*, and their chief function is as centers of exchange. These early cities of Sumer (lower Tigris-Euphrates valley), including Ur, Nippur, and later, Babylon, and the Nilotic and Indus cities were in some ways, nearly 6000 years ago, very like modern cities: houses crowded together, large public buildings and parks, familiar problems of congestion, sewage disposal, bureaucratic corruption, and interpersonal competition. They included, like modern cities, a wide range of occupations and specializations: merchants, officials, soldiers, priests, tailors, weavers, potters, carpenters, scribes, engineers, architects, poets, actors, masons, metal workers, and even bankers, lawyers, and doctors. By about 2000 BC similar urban developments were taking place in north China, and by at least 1000 BC in Mediterranean centers like Crete, Phoenecia (modern Lebanon), and southern Greece.

With the rise of the Roman empire and its contemporary in the East, the Han dynasty of China, very large and productive areas were unified under one rule and within one system of exchange. Rome, close to the center of the Mediterranean Basin and acting both as imperial capital and as an economic central place, probably reached a million in population by about the first century AD, aided by its coastal location so that it could bring in food and exchange trade goods by cheap sea transport. The Han capital at Ch'ang An, also probably with a million people, was on the Wei River (a tributary of the Yellow River) and depended in addition on a network of canals. The later Chinese capital at Hangchow, visited by Marco Polo in the thirteenth century, was even larger, in part because it was still better supplied by an intricate system of natural and artificial waterways, including the Grand Canal, near the coast of central China and in an even more productive agricultural area. In the West, the break-up of the Roman empire and the disruption of its trade routes after 450 AD was followed by a sharp decline in city growth. Rome itself, no longer maintained by its former tributary area, probably shrank to no more than 20,000 people and many of its monumental buildings were broken up for peasant huts. Trade in the West began to revive after about 1000 AD, further stimulated by the Crusades and their renewed contact with the eastern shores of the Mediterranean. Urban leadership shifted to Venice, whose location at the head of the Adriatic Sea meant that it was the closest Mediterranean seaport to northern Europe (via passes over the Alps) and at the same time closest by cheap sea routes to the Levant (eastern Mediterranean) and the East. In Marco Polo's time and after, Venice was the largest and richest city, as a center of inter-regional trade, in all of the West, although it still did not approach the size of Hangchow, the capital and chief central place of the much larger and richer Chinese empire.

With the revival of trade in Europe, and also after the European discovery of sea routes to the New World and to Asia, other cities there began to grow, the largest of them on coastal or riverine sites where access was maximized. There, in centers like Antwerp, Amsterdam, London, Hamburg, and Paris a new middle class of merchants began to emerge who became in time the most powerful groups within the new national states of Europe, displacing the older aristocracy based on ownership of land. It was in these cities that the outlines of the modern urban-centered world were established, and there that most modern institutions: commercial, political, and social were formed. The coming of industrialization and railways greatly accelerated this process, in part because it still further increased accessibility. Cities now could spread into all areas where there was a market to serve, and they could include a larger and larger share of the total population, now composed of industrial workers as well as merchants, bankers, artisans, and professionals. By the late nineteenth century the United Kingdom (Britain) became the first predominantly urban country, soon followed by the rest of Europe, and later by the United States, Russia, Canada, Australia, Japan, and other industrializing systems. In 1790 however only about 5 per cent of Americans lived in urban places, and until the coming of the railways and the boom in city growth after the Civil War, large cities were confined to the eastern seaboard, plus a few rapidly growing settlements along the Mississippi-Ohio system and the shores of the Great Lakes. As railways filled in the intervening spaces, new urban central places began to emerge and to extend their tributary areas as access improved, while at the same time supporting new industrial functions. By the census of 1920, over half of Americans lived in cities.

Definitional problems

Only in those parts of the world where agriculture is still the dominant sector of the economy and where most people are farmers (most of Asia and Africa, parts of Latin America), are city dwellers in a minority, although even there urbanization is increasing, as industrialization and transport routes spread. Precise definitions of "urban" are difficult, especially for settlements of smaller size: how to draw a line between a peasant village, a market town for farmers, and a city or town whose chief functions are non-agricultural? Different national censuses use different definitions of "urban," and some include settlements as small as 2000 in population, as does the United States Census, but in general the emphasis is on function: places or populations whose chief function is not *primary* (agriculture, mining, fishing, forestry, and other *extractive* industries) but *secondary* (manufacturing, construction) or *tertiary* (the service industries). The idea here is that primary industries produce basic raw materials and food, secondary industries process them or turn them into finished products, and tertiary industries exchange these things (wholesale and retail trade), finance the process (banking, insurance), administer, regulate, and tax them (government), and provide personal services (medicine, education, lodging, entertainment, cleaning, some government services, and so on).

Before industrialization most people everywhere provided most of these secondary and tertiary services for themselves, or used a neighbor's help; they processed their own crops and other raw materials, baked their own bread, built their own houses, treated their own sick, and so on, with only a relatively small role played by trade and merchants, by government, or by specialized industries such as grain milling, metal working, ship building or apprentice-style education. The process of economic development has really consisted of ever-increasing *specialization*, which, with the help of new power sources, is able to produce far more, and at lower cost, and at the same time to distribute it far more widely. By its nature, this process of specialization and *exchange* has always centered in cities; as it grew, so did the urban component. This trend is still increasing. In most of the world, cities continue to grow both more numerous and larger and to include a growing share of the population. As recently as 1900 there were only about a dozen cities in the world with populations of a million or more; by 1980 there were over 200.

Estimating the true size of a city's population presents another problem, since it is not easy to draw a distinct line around almost any urban area. Especially in industrialized countries with good transportation facilities, cities have spread out over large areas *(urban sprawl)*, often engulfing what earlier were separate settlements but also including areas which particularly at the fringes are ambiguous. These are sometimes called "rurban" (rural-urban) areas because they belong to some extent to both worlds. Cities are also dynamic entities, always in the process of change, so that their limits change too, expanding, retracting, or forming different patterns. Some urban characteristics or functions will also cover certain areas, other characteristics or functions different areas. As pointed out in Chapter 3, the political limits of a city seldom offer a useful guide, particularly in the United States, where most cities have long since outgrown their political boundaries or in any case no longer conform closely to them. The United States Census uses a much more effective basis for defining urban areas, the *Standard Metropolitan Statistical Area* (SMSA). The SMSA includes one (or more) cities of 50,000 population or greater, plus adjoining counties that have more than 75 per cent of their populations employed outside of agriculture. This appropriately emphasizes the distinctive nature of cities, not merely as concentrations of people but as functional centers of secondary and tertiary activity.

Types and sizes of cities

Most large cities combine nearly all urban functions; they are centers of trade, administration, a variety of specialized services, manufacturing, entertainment, culture, research, religion, education, medicine, and above all, of transport, since cities are pre-eminently based on access. Most cities are, however, specialized mainly in one set of functions, and some are quite sharply specialized. Planned national capitals are perhaps the best example: Washington, most United States state capitals, Canberra (Australia), Brasilia (Brasil), or Islamabad (Pakistan), all of which are nearly exclusively devoted to government except for essential local service industries. Statistically, the largest import into Washington is paper, and the largest export waste paper. Other cities are equally heavily specialized in manufacturing: Detroit, Akron, Pittsburgh, or Essen in the German Ruhr district. Others still are mainly regional commercial and financial centers: Denver, Dallas, or Atlanta. There are also urban centers whose chief function is recreation or tourism (Miami, Las Vegas), mining (Duluth, Butte), education (Chapel Hill, N.C., Urbana, Ill.) or transport (Sault Ste Marie, Panama City). Older cities, and most of the biggest cities, tend to be more diversified and to be significantly involved in nearly all urban functions, as with most of the big cities of east coastal United States. New York, by far the biggest United States city and probably second only to Tokyo in its metropolitan area population, is however not a major heavy manufacturing center. It is specialized instead in clothing and fashion industries, publishing, banking and finance, entertainment, trade and transport, and in all these functions serves as the central place for most of North America.

In many countries, including the United States, city sizes are not only distributed along a hierarchy, as pointed out in Chapter 3, but are related to one another according to the *rank-size rule*. This holds that if all cities within the national unit are ranked according to population size, multiplying the population of any one city by the number of its place in the rank will equal the population of the largest city, in rank 1. Thus for example, Philadelphia, rank 4, is roughly one-fourth the size of New York, rank 1, using SMSA figures, and Binghamton, N.Y., rank 84, is about 1/84th. The rank-size rule is however disrupted in many countries, especially many of the smaller ones, by the phenomenon of the *primate city*, where one city is over twice (or more) the size of its nearest domestic rival. This is especially striking in the cases of Denmark, where Copenhagen has over a quarter of the total Danish population, Greece (Athens), or Austria (Vienna), but is observable to varying degrees in a great many smaller countries where one outstanding location has come to dominate the urban hierarchy, especially as transport has improved. One might argue that even in the United States, what is sometimes called "the Bos.-Wash. Corridor" is becoming almost a single city, known even to the United States Census officially as *Megalopolis* (literally, "big city"). This is certainly by far the biggest urban concentration in the country, and while it does not operate entirely as a unit, in many respects it does function as a primate mass. Megalopolis, including Boston, Providence, New Haven, New York, Newark, Trenton, Philadelphia, Wilmington, Baltimore, and Washington, occupies about 2 per cent of the national space but holds about 20 per cent of national population, in a corridor which is almost unbrokenly urban.

Urban-based manufacturing

We have already discussed in earlier chapters why cities grow and why they arise in particular places. Manufacturing, the principal *basic* element in modern city growth (see Chapter 3), becomes more and more market-oriented as transport improves and as specialized labor, technical and managerial skills, and support services such as finance, research, or satellite speciality industries continue to increase their share in the total costs of manufacturing. These things are most readily and cheaply available in cities, most of all in the largest cities, while raw materials can be transported to and finished products from cities more easily. Power, in various forms, is also usually cheapest and most abundant in cities. Counter trends which make especially the larger cities less attractive to manufacturing include higher taxation, high costs of space (rent), the difficulty of atttracting skilled technicians or labor in some central urban locations, and (overdue) restrictions on industrial pollution. Large space users, high technology industries, and heavily polluting industries (such as petrochemicals) have in many cases tended to move out to the urban fringes where space is cheaper, skilled workers more easily recruited, or pollution can be concentrated in industrial zones separated from residential or commercial areas.

In many cases, such developments have formed new *satellite cities*, as much of urban northern New Jersey is to New York, or Gary, Indiana is to Chicago. Cleaner and higher-skilled industries, such as publishing or electronics, may often be found in satellite or suburban clusters in or near higher income residential areas. The spread of freeways and other roads and the increase in truck service has made such locations almost as central (accessible) as the middle of large cities, if not more so. In other cases, improved transport plus local amenities and lower local costs have combined to locate some manufacturing in previously rural or small town sites, but in the great majority of cases these are still within easy range of a major city. In many older American cities which grew originally along a navigable waterway or later around a railroad terminal, the old urban center has decayed as a result of such trends, further accelerated by the flight of higher-income residents to the suburbs, where pleasanter living conditions became easily accessible with the development of commuter lines and the private car. Cities have lost revenue to the suburbs, and many are increasingly unable to cope with the mounting problems of urban poverty, deteriorating services, and visual blight. In a sense, this is merely the result of the arbitrary and now outmoded administrative distinction between the city proper and its suburban and satellite fringes, where most new growth has been concentrated but which is usually politically separate. The city proper may be losing population, as most big United States cities are, but the metropolitan area is still growing in most cases; it is merely shifting its spatial pattern and emphasis. It is still a process of concentration, only slightly less exclusively pinpointed in one small site.

The center and the suburbs

Nevertheless, a *central business district*, or CBD, is still clearly discernible in nearly all cities, and still dominant in most of them. This is the central area of maximum accessibility by high speed transport from all outlying areas; it is marked by the largest single cluster of department stores, office buildings, government buildings, banks, theatres, hotels, and a variety of retail stores. Manufacturing is usually absent at the core of the CBD, where land values and rents are high, but some plants, especially in light industries, may be located adjacent to it. Los Angeles is a qualified exception to the pattern of a single dominant CBD, as are some of the newer cities in Texas and the Southwest, since all grew up almost entirely in the age of the private car and have hence developed a number of spatially scattered centers rather than one which is strongly dominant. Even in such cities, however, there is still a primary CBD as well as connections between it and subsidiary centers, all tied together effectively by high-speed freeways. Chicago, as illustrated in Map 5, may be more typical of most American cities in the pattern of commuting to a single dominant CBD from residential and suburban areas at varying distances, using both public mass transport (trains, buses, subways) and private cars.

As cities continue to grow outward, so may urban blight in the form of abandoned or dilapidated buildings in areas of low income and high unemployment where there are no longer resources to maintain structures or to build new ones and where crime and despair seem to dominate the landscape. This has become true for much of the northern edge of New York City (Harlem, the Bronx), of the south side of Chicago, and of similar areas in most big American cities. These are also the areas where minority groups, especially Blacks and Hispanics, tend to concentrate, since land values and rents are low. This is perhaps the greatest single failure of United States cities; they have created one of the richest and most productive economies in the world, but have not been able (often it seems have not cared) to provide a decent life for millions of their people, who have been left behind in the march of "progress" – outward to the high-income suburbs, the high-technology industries, and "the good life." The financial burden imposed by even minimal aid for these disadvantaged groups has strained and exceeded the means of most big

United States cities, and the solution is not easy. Employability and high incomes are more and more dependent on education and training, plus mobility, all advantages which minority groups mainly lack and find it increasingly difficult to obtain.

Ironically, efforts at revitalizing the central areas of American cities have concentrated mainly on the older downtown areas rather than on the decaying fringes around them. The reason of course is that this has been found to pay, and that there is development money available for it: new office and public buildings, new pedestrian malls and parks, and new or restored high-quality housing close to the CBD, often in the earlier heart of the city which has since been by-passed by newer commercial development but which is still convenient and even stylish residential space for high-income business and professional workers in the CBD. Many older cities have seen some revival of CBD functions, a trend which may continue as some high income residents move back to restored central city housing and a's exploding fuel costs restrict further suburban and *ex-urban* sprawl, outlying shopping centers, and other spatial expansion dependent on low cost transport by private car. Such a pattern of new investment in downtown or near-downtown office, business, and high-price residential development is apparent in New York, Chicago, Philadelphia, Boston, Baltimore, and many other older cities. Meanwhile the outer fringes of the CBDs in these and other cities, on the edges of the warehousing, wholesaling, and light manufacturing districts which often surround the CBD, continue to decay as slums for the far more numerous poor and disadvantaged. At the same time, suburban shopping centers, catering to other high-income "ex-urbanites," have expanded their functions to include not only retailing and major department stores but theaters, banks, legal, medical, and dental offices and clinics, car repair garages, and almost the full range of goods and services available in the CBD. One possible danger signal here is that the space occupied by parking lots in such shopping centers is many times greater than that occupied by buildings. The same is true of manufacturing plants and offices located in suburban "industrial parks" and "office parks," attractively landscaped but overwhelmingly dependent on the private car.

National urban patterns

In 1981 over three-quarters of the United States population lived in urban areas. Among them, United States cities effectively divide the national space into more or less discrete tributary areas, as in Map 29, based on newspaper circulation. However, urbanism is unevenly distributed, with the largest numbers of cities, and the biggest ones, as well as the higher urbanization levels, concentrated along the northeastern seaboard and the southern margins of the Great Lakes, with smaller clusters in central and southern California, the far corner of the Pacific Northwest (Portland and Seattle-Tacoma), and central-southern Texas. These are all areas

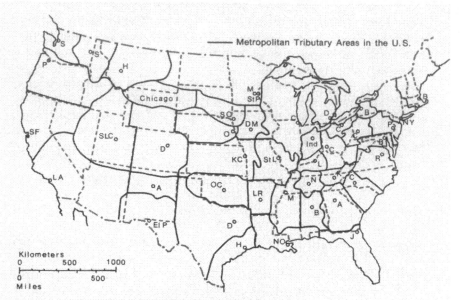

Map 29 Metropolitan tributary areas in the United States. *Areas covered by newspaper circulation from the dominant city in each area, indicated by a circle and an initial in each case. Can you name each city? Data are for 1980.*

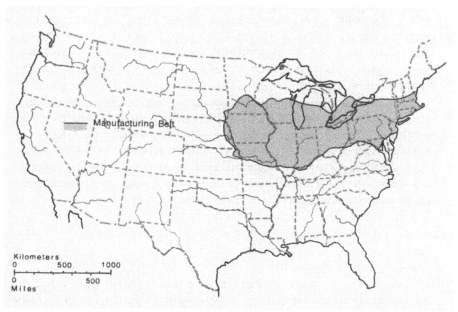

Map 30 The United States manufacturing belt. *Data are for 1980, a composite, of value added by manufacture and numbers of industrial workers.*

where manufacturing is prominent and where it continues to find strong advantages. Map 30, showing the manufacturing belt, demonstrates the close association with city growth, and further emphasizes the high degree of spatial concentration involved. Nearly half of total value added by manufacturing in the United States and about half of total industrial employment is in this manufacturing belt, which is also maximally accessible by transport lines to the rest of the country, including the waterways (coastal and inland) which were the original basis of its industrial and urban leadership. Maps 12 and 13 showing United States railroads and population distribution further emphasize this point.

Such major patterns change only slowly, but especially since the Second World War change has become more rapid. New industrial and urban developments in Texas, on the Pacific coast, and in the Old South, among others, have grown faster than the major cities and industrial centers of the main manufacturing belt. Americans have become increasingly mobile in search of attractive employment, and as incomes have also risen more people are moving to what are seen as environmentally attractive areas, especially the "sun belt" of California and the Southwest, Texas, Florida, and the rest of the South. By census records, the average American moves fourteen times in his or her lifetime! They are also an ageing population, with a higher proportion of older people, who with retirement benefits or higher incomes can afford to pursue area amenities. The sun belt is the nation's fastest growing area, including its cities. If and as solar energy becomes more important, these areas will get a further boost; already they are profiting from skyrocketing fuel costs, which bear far less heavily on them than on colder areas. In addition, energy-rich areas such as Texas (oil and gas) benefit from eager national demand and steeply rising prices.

The biggest cities tend however to have a certain self-perpetuating quality, as pointed out in Chapter 3, especially since they can perform functions or offer services found nowhere else, and can usually also offer the widest range of goods and services in great variety and with extreme specialization. New York is not seriously challenged as the country's largest metropolitan center, nor is its leadership in banking, finance, investment, foreign trade, publishing, or fashion, even though newer regional centers may increasingly compete with it in these and other respects: California and Florida in dress fashions, Texas or Chicago in finance, California ports in foreign trade, and so on. The biggest cities, with the greatest range of specialized functions and services, are often the most attractive locations for many other commercial, industrial, or service activities, who find there everything they need, and who are also thus placed at the center of the largest markets. What is true for New York tends to be true of other large metropolitan centers as opposed to smaller places.

Problems of urban size

But size also imposes its own problems: crowding, congestion, slowing of intra-urban movement, shortages of basic urban services (or a decline in their quality), rising rents and other costs, all of which are reflected in lowered efficiency and less desirable living or working conditions. There are also increasingly unmanageable problems with waste disposal and with air pollution, imposing high costs and destructive environmental consequences. Urban water demands, for consumption, industrial use, and waste disposal, have been shown as a mounting problem in Chapter 11. Big city growth in most areas is already hampered by water shortages, and by competition between areas for water as a vital resource. To depend on local ground water, by pumping from wells, may also have serious consequences; parts of downtown Houston have sunk over four feet in the past thirty years as underground aquifers have collapsed; the same problems have become even more serious in Mexico City, which is built on a former lake bed drained by the Spanish. Big city traffic congestion, in this land of the private car and truck, is not only wasteful, irritating, and expensive but a major cause of air pollution with even more serious consequences. And slowing of movement within or in the approaches to big cities strikes at the basis for all urban growth: *access*. If that is no longer easy, quick, and cheap, relatively speaking, the city loses competitiveness.

Access is of course also vital to the survival, let alone the dominance, of the CBD. The bigger a city grows, outward from its CBD, the more difficult it may become to reach the center from much of the urban area. This is of course the basis for the growth of suburban shopping centers, centers of service, and outlying industrial and commercial areas of all sorts. Earlier, the CBD area in most American cities depended primarily on rapid transit lines: subways, street cars, and commuter rail lines, which can move much larger volumes of people at greater speed than is possible by private car. As private car use expanded, especially after the Second World War, public transport in many cities was allowed to decay, or did not keep up with growing urban population. The CBDs have suffered, but as suggested already their decline may not be a permanent trend now that rising fuel costs pose a stronger disincentive to unlimited use of the private car. The decline in land values and in concentrated commercial functions outward from the CBD was never completely smooth however, as pointed out in Chapter 4. Railways, subways, and street car lines gave most cities a roughly star-shaped pattern of outward growth. Trucks and the private car later filled in many of the less developed areas between the points of the star, but at each period there were and still are subsidiary centers of relative accessibility and hence of functions like those of the CBD: major street or highway intersections, major rail or subway junctions, and on the outer fringes, freeway junctions where the biggest multi-purpose suburban shopping centers tend to locate.

Competition for space and the high cost of land in central urban areas began about the beginning of the present century to stimulate taller buildings, eventually skyscrapers, which now mark the CBDs of nearly all American cities. Since the Second World War there has been a boom in such high-rise structures, which have also spread to central city areas, in many other countries, as in London, Tokyo, Moscow, or Mexico City, responding to the same space pressures and high land values. Some of the larger high-rise buildings in American and other cities have developed into largely self-contained communities, providing a great variety of services in addition to offices: department stores, apartments, restaurants, hairdressers, banks, dentists, other personal services, a variety of smaller retail stores, and even internal parking space. In addition to vertical expansion, the core areas of big cities have also expanded underground, not only for the maze of power, sewer, and other utility lines and subways but for parking areas, underground shopping malls, and service facilities in the same variety as in the big high-rise buildings.

However, most new urban growth in most American cities is still around the edges, for residential, commercial, and industrial use. In 1980 nearly 40 per cent of the United States population lived in suburban areas, continuing a trend which began in the last decades of the nineteenth century as rail and street car lines began to make possible the "flight to the suburbs." While this has made serious financial problems for most cities, as already mentioned, urban renewal campaigns have concentrated on revitalizing the CBD. Central urban areas are now more attractive, but not appropriate, or even practicable, for mass residence, especially by lower-income groups. New access highways or freeways have been built in an effort to keep central urban areas accessible to higher-income suburbanites. In effect, the original city, its center, is moving toward more complete specialization in commerce, services, and service industries, while the outer areas take over more of the residential function and at the same time provide space for industrial expansion and for local centers of trade and business. Such trends do not mean the decline, still less the "death," of cities, as one sometimes hears said, but rather a rational change in spatial form and in areal specialization. The confusion is not helped by the arbitrary political lines which divide the city proper from its suburban areas; both parts are in fact *symbiotic* (mutually interdependent) partners, and could not exist without each other; they belong to a single functional whole. Most suburban residents in most cities everywhere still commute to employment within the city proper, even though a rising proportion now find jobs within the suburban area and its own range of industrial and service functions.

It is part of the basic nature of all cities that they are *dynamic* entities, always in the process of change. They have been and remain the chief centers of change within every society, and that role is reflected within them. The city is in fact a mirror of the society which builds it, as well as a shaper of that society. This was true in the pre-industrial past, when cities represented the pinnacle of each society's sophisticated culture and the chief centers of activity and ferment, even when they included only a small minority of the total population. Now that expanded urban areas increasingly hold a majority of the population, cities are still more the foci of what each society represents, and of where it is going. To study the city is to study ourselves. Since most of us now live in an urban world, it is perhaps an appropriate note on which to end this book: try to use what you have learned from it, and look around you!

SUGGESTIONS FOR FURTHER READING

Chapter 1

Chapman, K. *People, Patterns, and Processes: An Introduction to Human Geography*. Wiley, 1980.
Getis, A. and J., and Fellman, J. *Geography*. Macmillan, 1981.
Haggett, P. *Geography: A Modern Synthesis*. 3rd ed. Harper & Row, 1979.
James, P. E., and Martin, F. J. *All Possible Worlds: A History of Geographic Ideas*. 2nd ed. Wiley, 1981.
Jordan, R., and Rowntree, L. *The Human Mosaic: An Introduction to Cultural Geography*. 2nd ed. Harper & Row, 1979.

Chapter 2

Birdsall, S. S., and Florin, J. W. *Regional Landscapes of the United States and Canada*. 2nd ed. Wiley, 1981.
Healey, R. G., and Rosenberg, J. S. *Land Use and the States*. Johns Hopkins Press, 1976.
Jackson, R. H. *Land Use in America*. Wiley, 1981.
Weinsein, B. L., and Firestine, R. E. *Regional Growth and Decline in the United States*. Praeger, New York, 1978.
Whittlesey, D. S. "The Regional Concept and the Regional Method," in P. E. James and C. F. Jones, eds, *American Geography: Inventory and Prospect*. 2nd ed. Syracuse, 1964.
Wilson, A. G. *Urban and Regional Models in Geography and Planning*. Wiley, 1974.
Zelinsky, W. *Cultural Geography of the United States*. Prentice-Hall, 1973.

Chapter 3

Downs, R. M., and Stea, D. *Maps in Mind*. Harper & Row, 1977.
Gould, P., and White, R. *Mental Maps*. Penguin, Baltimore, 1974.
Morrill, R. L., and Dormitzer, J. M. *The Spatial Order: An Introduction to Modern Geography*. Duxbury Press, Scituate, 1979.
Muehrcke, P. C. *Map Use: Reading, Analysis, and Interpretation*. J. P. Press, Madison, 1978.

Robinson, A. H., and Petchenik, B. *The Nature of Maps*. Chicago, 1976.
Taafe, E. J., and Gauthier, H. L. *Geography of Transportation*. Prentice-Hall, 1973.

Chapter 4

Brown, L. A. *Innovation Diffusion*. Methuen, 1981.
Herbert, D. T., and Johnston, R. S. *Social Areas in Cities*. Wiley, 1978.
Lowe, J., and Moryadas, S. *The Geography of Movement*. Houghton Mifflin, Boston, 1975.
McNeil, W. H., and Adams, R. S. *Human Migration: Patterns and Policies*. Indiana, 1978.
Smith, D. M. *Geographic Perspectives in Inequality*. Barnes & Noble, New York, 1979.
White, P. E., and Woods, R. *The Geographical Impact of Migration*. Longmans, New York, 1980.

Chapter 5

Bennett, C. F. *Man and Earth's Ecosystems*. Wiley, 1975.
Blouett, B. W., and Luebke, F. C., eds. *The Great Plains: Environment and Culture*. Univ. of Nebraska Press, Lincoln, 1979.
Dasman, F. R. *Environmental Conservation*. 4th ed. Wiley, 1976.
Hammond, K. A., Macinko, G., and Fairchild, W. B., eds. *Sourcebook on the Environment*. Univ. of Chicago Press, 1978.
Kneese, A. V. *Economics and the Environment*. Penguin, New York, 1977.

Chapter 6

Burton, I., *et al. The Environment as Hazard*. Oxford, New York, 1978.
Foin, T. C. *Ecological Systems and the Environment*. Houghton Mifflin, Boston, 1976.
Freeman, A. M. *The Benefits of Environmental Improvement*. Johns Hopkins Press, 1979.
Kneese, A. V., and Bower, B. T. *Integrated Environmental Quality Management*. Johns Hopkins Press, 1979.
Miller, G. T. *Living in the Environment*. 2nd ed. Wadsworth, Belmont, California, 1979.
Owen, W. *Transportation in Cities*. Washington, 1976.
Richardson, J. L. *Dimensions of Ecology*. Williams & Williams, Baltimore, 1977.
Stoker, H. S., and Spencer, L. S. *Environmental Chemistry: Air and Water Pollution*. 2nd ed. Glenview, Ill., 1976.

Chapter 7

Drury, G. H. *Introduction to Environmental Systems*. Heinemann, Exeter, N.H., 1981.
Gardner, J. S. *Physical Geography*. Harper & Row, 1979.
King, C. A. M. *Physical Geography*. Barnes and Noble, New York, 1980.
McIntyre, M. P. *Physical Geography*. 3rd ed. Wiley, 1980.
Strahler, A. N. and A. H. *Elements of Physical Geography*. 2nd ed. Wiley, 1979.

Chapter 8

Boucher, K. *Global Climate*. Wiley, 1975.
Gaskell, T. F., and Morris, M. *World Climate*. Thames & Hudson, 1979.

Griffin, J., ed. *Climatic Change*. Cambridge Univ. Press, 1978.
Parry, M. L. *Climatic Change, Agriculture, and Settlement*. Shoestring Press, Hamden, Conn., 1978.
Trewartha, G. T., and Horn, L. H. *An Introduction to Climate*. 5th ed. McGraw-Hill, 1980.

Chapter 9

Adams, R. and M., and Willens, A. and A. *Dry Lands: Man and Plants*. St Martins Press, New York, 1979.
Clawson, M. *Forests for Whom and for What?* Johns Hopkins Press, 1975.
Eyre, S. R. *Vegetation and Soils*. 2nd ed. Aldine, 1968.
Walter, H. *Vegetation of the Earth*. London, 1973.

Chapter 10

Foth, H. D., and Turk, L. M. *Fundamentals of Soils Science*. 5th ed. Wiley, 1972.
Kirkby, M. J,, ed. *Soil Erosion*. Wiley, 1981.
Steila, D. *The Geography of Soils*. Prentice-Hall, 1976.
Symons, L. *Agricultural Geography*. 2nd ed. Westview, Boulder, 1978.

Chapter 11

Bowen, R. *Ground Water*. Wiley, 1980.
Hunt, C. A., and Garrells, R. M. *Water: The Web of Life*. New York, 1972.
Leopold, L. B. *Water: A Primer*. Freeman, San Francisco, 1974.
Lieber, H. *Federalism and Clean Water*. Lexington, Mass., 1975.
McCaull, J., and Crossland, J. *Water Pollution*. New York, 1974.
Smith, D. I., and Stopp, P. *The River Basin: An Introduction to Hydrology*. Cambridge Univ. Press, 1978.
White, G. F. *Strategies of American Water Management*. Univ. of Michigan Press, Ann Arbor, 1969.

Chapter 12

Butzer, K. W. *Geomorphology from the Earth*. Harper & Row, 1976.
Embleton, C., and Thornes, J., eds. *Process in Geomorphology*. Wiley, 1979.
Muller, R. A. *et al. Physical Geography Today*. 2nd ed. Random House, 1978.
Press, F., and Siever, R. *Earth*. Freeman, San Francisco, 1976.

Chapter 13

Haup, A., and Kane, T. T. *The Population Handbook*, Washington, D.C., 1978.
Hornby, W. F., and Jones, M. *An Introduction to Population Geography*. Cambridge Univ. Press, 1980.
Peters, G., and Larkin, R. *Population Geography: Problems, Concepts, and Prospects*. Kendall/Hunt, Dubuque, Iowa, 1979.
Repetto, R. G. *Economic Equality and Fertility in Developing Countries*. Johns Hopkins Press, 1979.
Skinner, B. J. *Earth Resources*. 2nd ed. Prentice-Hall, 1976.
Tapinos, G. and Piotrow, P. *Six Billion People*. McGraw-Hill, 1978.

Chapter 14

Anell, L., and Nygren, B. *The Developing Countries and the World Economic Order*. Methuen, 1980.

Cook, E. *Energy and Society*. Freeman, San Francisco, 1976.

Darmstadter, J. *et al. Energy in America's Future*. Johns Hopkins Press, 1979.

Dunkerley, J., ed. *International Comparisons of Energy Consumption*. Resources for the Future, Washington, D.C., 1978.

Halacy, D. *Earth, Water, Wind, and Sun: Our Energy Alternatives*. Harper & Row, 1977.

Odum, H. T., and E. C. *Energy: Basis for Man and Nature*. McGraw-Hill, 1976.

Smith, W. K., ed. *Scarcity and Growth Reconsidered*. Johns Hopkins Press, 1979.

Wilson, C. L., ed. *Energy: Global Prospects 1985–2000*. New York, 1977.

Chapter 15

Barlowe, R. *Land Resource Economics*. 2nd ed. Prentice-Hall, 1972.

Hamilton, F. E., ed. *Contemporary Industrialization: Spatial Analysis and Regional Development*. Longmans, New York, 1978.

Hamilton, F. E., ed. *Spatial Perspectives on Industrial Organization and Decision Making*. Wiley, 1974.

Hitch, C. J., ed. *Resources for an Uncertain Future*. Johns Hopkins Press, 1978.

Miller, E. W. *Manufacturing: A Study of Industrial Location*. Penn State Press, 1977.

Park, C. F. *Earthbound: Minerals, Energy, and Man's Future*. Freeman, San Francisco, 1975.

Wagner, R. H. *Environment and Man*. Norton, New York, 1978.

Chapter 16

Abu-Lughod, J., and Hay, R. *Third World Urbanization*. Methuen, 1980.

Berry, B., and Gillard, Q. *The Changing Shape of Metropolitan America*. Ballinger, Cambridge, Mass., 1977.

Carter, H. *The Study of Urban Geography*. 2nd ed. Arnold, London, 1976.

Hartshorn, T. A. *Interpreting the City: An Urban Geography*. Wiley, 1980.

Murphey, R. *The Fading of the Maoist Vision: City and Country in China's Development*. Methuen, 1980.

Northam, R. M. *Urban Geography*. 2nd ed. Wiley, 1979.

Sternlieh, G., and Hughes, J. W. *Shopping Centers USA*. Rutgers Press, 1980.

Yeates, M., and Garner, B. *The North American City*. 2nd ed. Harper & Row, 1980.

POSSIBLE STUDY/DISCUSSION QUESTIONS

Chapter 1

1. Why is it reasonable to say that "All geography begins with maps"?
2. How and why does social science differ from physical science?
3. If everything has a cause, why is it often so hard to discover causes?

Chapter 2

1. Divide North America into regions at three different orders of magnitude. Characterize each region, define its core, and draw boundaries around it. What kinds of problems arise over transition zones or zones of indifference?
2. What characteristics or factors, and in what order of importance, do you find it necessary or useful to consider in making the above regional distinctions?
3. What kinds of single-factor or functional regions, and how many, would you distinguish in North America?
4. Find other examples of regional cores. How do they differ from each other?
5. Apply a three-level regional division to your own county or state, and then to the town where you live.
6. Select and discuss two examples of settlements whose present status must be explained primarily in terms of historical rather than current factors.

Chapters 3 and 4

1. What is the relationship between specialization, exchange, and economic development?
2. In what ways is urbanization dependent on spatial interaction?
3. Are there any aspects or characteristics of human society which do not have spatial form or which cannot be mapped? What does the answer suggest?
4. Select an example of spatial interaction within your city or town, within your county or state, and at the national level.
5. Measure on an atlas map the distance between (a) Los Angeles and Tokyo, (b) London and Mexico City. Compare both with the distances shown by measuring on a globe. How do you account for the differences? How are they related to the projection used in making the flat map you used?
6. Using an atlas map, find the cities at 40° N latitude, 75° W longitude; 4° 50′ N lat. 74° W long.; 34° S 151° E; 52° 40′ N 2° W.

7. How are longitude and latitude determined for any given position, at sea for example?
8. How does spatial interaction affect regional distinctions? How do regional, cultural or economic differences affect spatial interaction?
9. What factors affect a decision to migrate? What are some contemporary patterns of migration? How can each be explained?
10. Try to discover your own mental maps, or your own spatial perception, by attempting to draw a map of your town, county, state, or nation, including the location of major cities, rivers, and political boundaries, and then comparing it with a printed map of the same area.

Chapters 5 and 6

1. What examples can you find, in addition to those cited, of similar environments which are used differently? Of different environments used similarly? How can each case be accounted for?
2. What other examples can you find of the similar use of similar environments which help support the idea that the physical environment does exert an influence on human activity? (To begin with, try maps of world agriculture.)
3. What environmental factors are relevant in the site of the town where you live? In its economic specializations?
4. Compare Switzerland and Bhutan. Their physical environments are similar, and the countries are about the same size. How are they different, and why?
5. To what extent is it reasonable to regard geography as "human ecology"? Why or why not is such a label appropriate?
6. Has "man conquered nature"?
7. Deforestation has often been called the single most destructive human action against the environment. Why? Is this in fact a reasonable statement?
8. Can industrial development and sound ecological practice co-exist? How?

Chapters 7 and 8

1. Is the climate of your area marine or continental, or can it not be clearly labelled as one or the other? Why?
2. Why should a productive agricultural base be a factor in industrialization?
3. How does climate affect interregional exchange?
4. What is the length of the growing season in the area where you live? How does this affect settlement and the use of land and other resources?
5. Denver is said to have the largest hinterland in area of any city in the United States. Why? Why then is Denver not a larger city?
6. What is the predominant form of rainfall in hot deserts? Why? What is this likely to mean for human settlement?
7. What is meant by *micro* climate? How does it differ from *macro* climate?
8. What is a diurnal monsoon?
9. What specific detailed information would you want about the climate of an area in planning the development or extension of an agricultural system? Why would each piece of information be necessary?

Chapter 9

1. Where are the remaining major forest stands in the world? Which are currently being exploited commercially for lumber on a large scale? Why in each case, and why have other forest areas been less used?
2. How was natural vegetation related to regional differences in pre-Columbian Indian cultures in North America? To regional cultural differences in Roman Europe?
3. Why is vegetation such a key element in all river basin planning programs?

4. Can you find place names in your area to suggest that forest once covered the sites?
5. What plant and animal species have multiplied as a result of human activities? Why in each case?

Chapter 10

1. Where are the local market gardens which supply the city or town where you live? Why? Are the soils they are using part of the reason? Are other factors involved?
2. Why should land which can profitably grow crops which are physically demanding be expensive?
3. In what detailed respects is wheat growing in the steppe areas of North America or Australia *extensive*, as compared with *intensive* agriculture in New Jersey or Holland?
4. Why do most tropical soils tend to be poor and of low fertility?
5. What are the disadvantages of single-crop agriculture? Why is it so widely practiced? Is it for the same reasons in each case?

Chapter 11

1. Where is the power plant or plants in the city or town where you live? Why? Is water the only locating factor?
2. Why should there be especially complex disputes over the use of water from the Colorado River? Why should India and Pakistan have had conflicts over water use?
3. What is a "power grid"? How does it affect the possibilities of transmitting electric power?
4. What other examples can you find of confluence cities? bridge-site cities? cities at upstream heads of navigation? outports?
5. How can one explain the distribution of irrigated lands in the United States, as shown on Map 20? What does each major irrigated area grow?

Chapter 12

1. Is there any general pattern of city location which helps to explain the growth of cities near the lowland approaches to mountain passes (see Chapter 3)?
2. Give a description of the country shown in Map 4.
3. Construct a vertical profile from a topographic sheet.
4. Why are there fewer distinctively mountain cultures in North America than in Europe or Asia?
5. What was the origin of the "Whiskey Rebellion" of 1794 in the United States? How were the grievances of the rebels settled?

Chapter 13

1. What does *intervening opportunity* mean? Give examples, and discuss its relevance to regional development as well as to resource use.
2. What current changes in resource use, and what new resources, can you discern in the contemporary world?
3. What are the causes of famine?
4. What is the national population density (people per square mile or square kilometer) of the Netherlands, Java, Tibet, Australia, Canada, Malta, and Puerto Rico? Which of these countries is "overpopulated"? What does "overpopulation" mean?

Chapter 14

1. Is a policy of "no growth," as sometimes urged, a workable course for already "developed" economies, especially as a response to energy problems and to problems of pollution?
2. What should the "developing" countries do about industry-linked pollution in their own territories? (The answer is not an easy one.)
3. What are the prospects that we can significantly reduce, or remove, our dependence on or use of fossil fuels, in the next fifty years? In the next century?
4. If solar energy could become an economically competitive alternative to fossil fuels, how would this affect the global distribution of manufacturing?
5. Why are plants using steam-driven turbines to produce *thermal power* the largest source of electricity for most cities? Why are they used even in areas rich in hydro power?

Chapter 15

1. What other major raw materials lose much weight in the manufacturing process? How does this affect the location of the manufacturing in question? The transport of the raw material?
2. Compare the location of manufacturing (Map 26) with the world maps of coal, iron, and petroleum. What do you deduce from the comparison?
3. What articles or goods in daily use have no connection with steel? How does the supply of, for example, fresh fruit or clothing, depend on steel?
4. If all rocks or soils become economically usable sources of all minerals, how would this affect the locational pattern of manufacturing?
5. What is "wise resource use"? What is "misuse"?

Chapter 16

1. Find examples of cities which have declined, at least relatively, as manufacturing sites, and explain why in each case.
2. Does the town or city where you live fit any of the several model patterns of different zones of land use? Can you define the CBD? An inner-urban ring of decay around it? A process of outward movement of residences ("flight to the suburbs")? Districts of wholesaling, warehousing, transport facilities, retail store clusters, financial services, entertainment, outlying shopping centers, ethnic areas, or other distinctive forms of land use?
3. Describe briefly the patterns of urban land use posited by the *concentric circles*, *sector*, and *multiple nuclei* models of urban growth. How well do any of these models fit the place where you live?
4. Why did cities develop earlier in the Middle East, India, China, and the Mediterranean than in northern Europe and other areas now far more highly urbanized?
5. Why is small-scale local industry usually more expensive (in the unit costs of its output) than large urban plants?
6. Draw a line around the urban area where you live; around New York city; what kinds of problems or ambiguities do you encounter?

INDEX

acid rain, 200
agricultural revolution, 181
agriculture, 74–6, 128–46, 194–5
alcohol, 199
alluvium, 133, 142
Alps, 64–70
aluminium, 210–11
Appalachians, 141
aquaculture, 196
artesian, 149

basic-service ratio, 29
biogas, 198–9
biotic resources, 177
birth control, 186–7
birth rate, 183–8
boundaries, 8–22, 31
Brazil, 178–9
break-in-bulk, 26, 62

California, 27, 34, 77, 213
Canada, 110–11
Canadian Shield, 111
causality, 3–7
central business district (CBD), 238–9, 242
central place, 30
charcoal, 118
cheese, 66
China, 13–14, 82, 237
chinook, 101
chocolate, 66
city, 8, 28–33, 45–51, 60–3, 173–4, 229–48
Civil War (US), 141
climate, 90–5, 96–116, 125–6
climatic change, 92–5
climatic determinism, 94–5
coal, 178–9, 197, 200, 210, 216–33
coffee, 168

coke, 178
Columbia River, 155
commercial revolution, 181
comparative method, 27–8
confluence, 160
conservation, 227
continental climate, 101–4
contour, 163
contour ploughing, 136
contraception, 186–7
convectional precipitation, 99–100
core, 9, 14–16
corn, 144
corn belt, 144
cotton, 75, 136
cyclonic precipitation, 98–9

dairying, 65, 143
death rate, 183–8
deforestation, 74, 117–21, 123–6
Delaware River, 63
deltas, 142, 160–1
democracy, 69–70
Denver, 108
diffusion, 53–5, 126
disease, 54, 74
dung, 197
dust bowl, 125

ecology, defined, 71
economies of scale, 157, 179
effective climate, 125
energy, 86–7, 190–207
environment, defined, 57
environmental determinism, 6, 58–60
erosion, 124–6, 136
evaporation rates, 97
extensive production, 108, 121

fall line, 61–3
fallowing, 135
fertility (soil), 134–6
fertilizers, 74–6, 134–6
fire, 117–18, 125–6
fishing, 195–6
floods, 123–5
flow resource, 197
foehn, 101
forests, 74, 115–21
fossil fuels, 72, 87, 193
fruit, 85–6, 138
functional region, 17

garbage, 199
geothermal energy, 198
gestalt, 10
glaciation, 79, 133
goats, 58
Grand Coulee Dam, 155–7
grasslands, 121–3
grazing, 121–3
Great Plains (US), 109, 125
green manure, 138
Green Revolution, 194–5
ground water, 147–50
growing season, 97

Hangchow, 233
hazard, 80–1
hinterland, 25, 29–33
history, 28
humus, 130–1
hydroelectric power, 154–7, 197–8

indirect benefits, 157
industrial revolution, 181, 208
industrialization, 181, 208–28, 230–3
insolation, 101
intensive production, 108, 138
intervening opportunity, 178
iron, iron ore, 210–11, 216–23
irrigation, 74, 139, 153–4

Japan, 187

land use, 20
language, 11, 23, 69
laterite, 130
latitude, 40, 101
laws, 3–7
leaching, 130
legume, 135
levee, 125
life expectancy, 203–6, 232
line costs, 157
literacy, 185
living standards, 187–8
loess, 133
logging, 119–21
local relief, 162
London, 15–16, 77, 161
longitude, 40

Los Angeles, 152
lumbering, 119–21

Malthus, Thomas, 186
manufacturing, 165, 181, 213–21
manufacturing belt (US), 240–1
maps, 34–41, 162–3
maritime climate, 101–4
market gardening, 145–6
Marxism, 6
Mediterranean Basin, 33–4
Megalopolis, 236
mental maps, 45
migration, 52–4
milk, 65–6
mining, 221–3
monsoon, 104–6
Montreal, 160
mountains, 162–75
mountain passes, 169–70
mountain cultures, 171–2

natural gas, 197
Netherlands, 12
New England, 145
New Orleans, 142
New York City, 29, 31, 62, 152, 236
nomadism, 121–3
nuclear energy, 201–3
nucleated settlements, 23

oceans, 195–6
oil, 197, 201, 216–33
oil shale, 201
orographic precipitation, 100
outport, 161

paper, 120
Paris, 15–16
per capita incomes, 203–6
perception, 72–3, 78, 80–3
pesticides, 194–5
petroleum, *see* oil
Philadelphia, 63
plankton, 196
planning, 9, 20
plants, 116–27
podzol, 130–1
pollen, 79
pollution, 71–83, 152, 192, 194–5, 207, 225
population density, 23, 89, 180
population growth, 181–9
possibilism, 60
poultry, 143
power plants, 152
precipitation, 96–101
prediction, 37
projections, 39
pulp, 120
Pyrénées, 172

quantitative analysis, 3–7

race, 52, 56
railroads, 109–11
rain shadow, 101
rainfall, 96–101
rainforest, 114, 118–21
rank-size rule, 236
reclaimed land, 173–4
refuge areas, 118, 166
regions, 8–22
regionalism, 10–11
regional planning, 9, 20
resources, 176–89, 208–28
resource appraisals, 224–8
Rhine River, 12, 158
rice, 107, 137–8
rivers, 10, 112, 124–5, 133, 152, 158–61, 165
Robin Hood, 116
Rome, 233
rubber, 212

satellite cities, 237
Schuylkill River, 63
science, 1–7
scientific method, 4–6
seasons, 105–6
shifting cultivation, 135
shopping centers, 239
silt, 124–5, 133, 197
sites, 24, 173–4
situation, 24
slavery, 141–2
slums, 238–9
smog, 77
social science, 1–7
soil, 128–46
soil fertility, 134–6
soil energy, 193, 200
spatial interaction, 23–41
Spain, 172
Standard Metropolitan Statistical Area, 235
steel, 151, 210–11, 213, 219
steppe, 107, 121–3
suburbs, 237–9, 243
sun belt, 241
Switzerland, 64–70

taconite, 211
taiga, 57, 114, 198
tea, 168
telephones, 55
tenancy, 141
terminal costs, 157
territoriality, 43
"Third World", 203
Tibet, 167
tides, 198
timber, 119–21
Timbuktu, 24–7
Tokyo, 80
tourism, 67, 169
trade winds, 102
transhumance, 65
transport costs, 26, 120, 144, 157–61, 169–71
tree crops, 138
tropics, 106, 135, 137, 138
truck crops, 145–6

"urban", defined, 234–5
urban blight, 238–9
urban history, 232–4
urban planning, 20
urbanization, 229–44

vegetation, 116–27
Venice, 233–4
villages, 46–8

water, 147–61
water power, 154–7
water rights, 153
water table, 147–9, 154
water transport, 157–61
watersheds, 125
waterways, 158–61
weather, 96–113
wheat, 33, 143–4
whiskey, 167
wind power, 198
winds, 102–6
wood, 197

zoning, 20